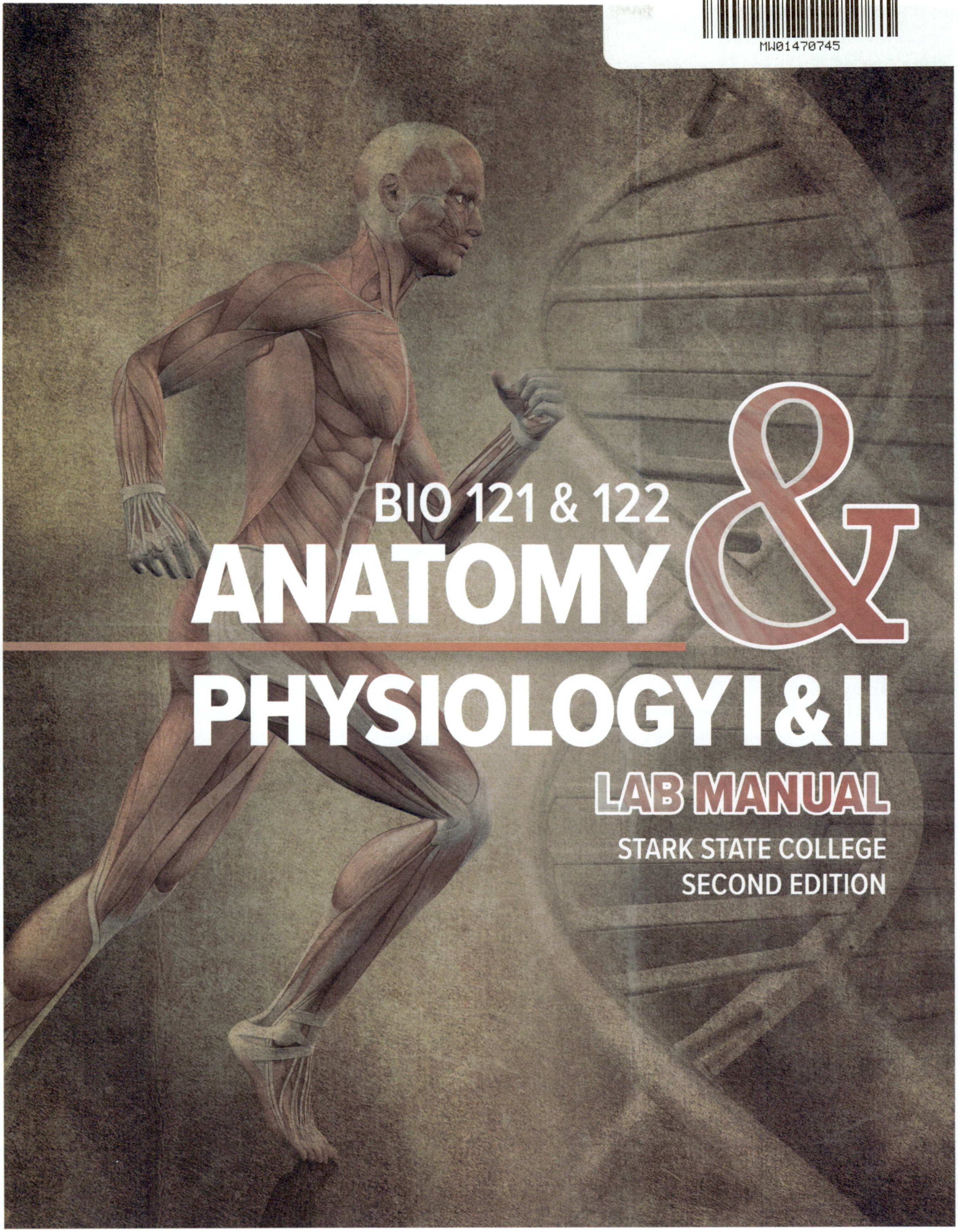

BIO 121 & 122

ANATOMY & PHYSIOLOGY I & II

LAB MANUAL

STARK STATE COLLEGE

SECOND EDITION

VAN-GRINER LEARNING

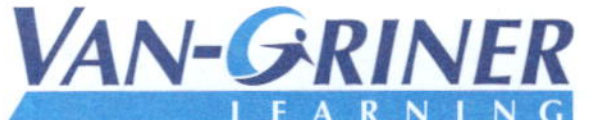

Anatomy and Physiology I and II, BIO 121 & 122

Laboratory Manual
Stark State College
Second Edition

Printed in the United States of America
10 9 8 7 6 5 4 3 2 1
ISBN: 978-1-64565-329-5

Van-Griner Learning
Cincinnati, Ohio
www.van-griner.com

President: Dreis Van Landuyt
Senior Project Manager: Maria Walterbusch
Customer Care Lead: Lauren Wendel

Tickner 65-329-5 Su23
331259

Table of Contents

While in the anatomy and physiology laboratories, the following safety guidelines ***must*** be followed at all times.

1 | No eating, drinking, or application of cosmetics (including lip balm) is allowed in the laboratories to prevent possible contamination.

2 | No children or individuals who are not Stark State anatomy and physiology students are allowed in the laboratories. Potentially hazardous materials and sharp instruments may be present in the lab which can pose safety concerns and liability issues.

3 | Follow only the authorized lab procedures and instructions as outlined in the laboratory manual or given by your lab instructor. If you have any questions concerning the lab procedures, please consult your instructor before proceeding.

4 | Please thoroughly clean up your lab station after each lab. This includes returning all lab materials and specimens to their proper locations, cleaning and drying any equipment, and properly disposing of any waste in locations as indicated by your lab instructor.

5 | Prior to leaving lab, please wash your hands in order to prevent the spread of communicable disease.

6 | Use caution when handling scalpels, other sharp equipment, and chemicals in order to prevent injury to yourself or others.

7 | Because of the expense of lab equipment and models, please treat them with caution and respect to prevent damage.

8 | Inform your instructor of any broken equipment or materials in the lab. If any glass or scalpels break, the instructor needs to properly dispose of the broken materials in the appropriate containers. In addition, the instructor needs to inform the laboratory manager for replacement and inventory of any broken or damaged lab materials/equipment.

9 | With the lab instructor's guidance, learn the locations and, if necessary, the operation of the following:

   a | Fire alarm

   b | Emergency exit route

   c | Fire extinguisher

   d | Fire blanket

   e | First aid kit

   f | Eyewash station and safety shower

   g | Automated external heart defibrillator (1st floor by elevator)

   h | Stark State security phone number (330-704-2582). We suggest that this number be programmed into your cell phone.

10 | Please report any lab accidents to the lab instructor. Accident reports may need to be processed for laboratory accidents.

11 | Physical examination and limited dissection of preserved anatomical specimens is an integral part of the laboratory experience. When participating in these lab activities, it is mandatory to wear gloves. Liquid resistant lab coats, goggles or safety glasses, and respiratory protection are optional but recommended. Wearing contact lenses may be inadvisable because they may trap irritants on the surface of the eye.

12 | It is the responsibility of the student to understand and evaluate the risks associated with participation in anatomy and physiology labs. Exposure to chemicals such as formaldehyde, phenols, and others is possible. Every anatomy and physiology lab has Safety Data Sheets (SDS) available which provide detailed information about any chemicals that may be present in the labs. If you would like a copy of the SDS sheets, your instructor can provide them to you as a complete packet.

13 | If you have any health concerns (such as asthma, allergies, pregnancy), please inform your instructor who will provide you with the SDS packet which you can share and discuss with your physician. You may be required to have a signed release from your physician in order to participate in the lab.

14 | For your safety, please use the following directions for sitting in the BioFit lab chairs. Prior to sitting, grasp the back of the chair so that it doesn't slip. Always sit so that your back is in contact with the chair back; do not sit on the edge of the seat. The weight limit for the chairs is 350 lbs. at 24/7 use. If this capacity does not meet your needs, you are encouraged to inform your instructor after lab to discuss other options to meet your needs. Such conversations will remain confidential.

I have read, understand, and agree to abide by this Safety Contract. ***Failure to adhere*** to the above regulations or engaging in any other lab conduct that is deemed hazardous by the lab instructor may result in my removal from the laboratory.

Date: ______________________________

Name (Signed): ______________________________

Name (Printed): ______________________________

Please keep this signed Safety Contract in your lab manual for future reference, if needed.

Cadaver Bill of Rights

At Stark State College, students and faculty are fortunate to use human cadavers to assist in the learning process in the anatomy labs. Out of respect for the generous gift of learning that these donors have given us, we have developed a Bill of Rights similar to that in many hospitals. Please read the following rights of the cadavers and sign the agreement at the bottom of the page indicating that you, as a student, will honor and respect their gift.

Stark State College recognizes that the cadavers have a number of basic rights. These include, but are not limited to, the following:

- The **right to considerate and respectful treatment** before, during, and after labs, at all times and under all circumstances, with recognition of their personal dignity.
- The **right to privacy,** both in the laboratory setting and outside of the laboratory. This right to privacy includes ***absolutely no photography*** (including by cell phones and computer tablets) at **any** time.
- The **right to expect confidentiality** which assumes that any discussion of the cadavers outside of the lecture or laboratory setting in public areas within or outside of the college will be conducted discreetly and privately.

I understand and agree to abide by the Cadaver Bill of Rights. Failure to adhere to these standards may result in my removal from the lab and disciplinary action according to Stark State's Policy and Procedures Manual.

Date: ______________________________

Name (Signed): ______________________________

Name (Printed): ______________________________

Please keep this signed Cadaver Bill of Rights in your lab manual for future reference, if needed.

1

The Scientific Method

Introduction

Scientific discoveries are often the result of one's curiosity about how something is organized or how it works. This curiosity drives scientific research to solve a problem or answer a question. In modern science and medicine, a standardized process called the **scientific method** is used to organize and evaluate information during scientific investigation in order to reach a logical conclusion. There are several steps to this process:

1 | **Identify the question or problem.** The question you are trying to answer or the problem you are trying to solve likely arose from observations of the world around you.

2 | **Formulate a hypothesis.** A hypothesis is an educated guess about the answer to your question or problem. It's more than a "hunch" or "gut feeling." It is an idea that may be formed from patterns seen in your observations, and that idea may or may not be correct. A hypothesis must be evaluated to determine if it is

- **Testable:** It can be studied through experimentation and data collection.
- **Repeatable:** Anyone who tests the hypothesis must obtain the same results. If the results vary, you must doubt the hypothesis.
- **Unbiased:** Bias may occur when the experimental data is not representative of the target population, leading to experimental error. There are many factors that could lead to bias, such as lifestyle, occupation, healthcare, socioeconomics, and population profile, to name a few!

As you try to solve a problem or answer a question, you should apply logical reasoning involving either **deductive** or **inductive reasoning.**

- Deductive reasoning involves two or more verified premises followed by a conclusion. For example:
    - **Assumption/Observation 1:** All cuts to the skin bleed.
    - **Assumption/Observation 2:** All cuts to the skin are painful.
    - **Assumption/Observation 3:** Susie has a cut in her skin.
    - **Conclusion:** Susie is bleeding and she is in pain.

- Inductive reasoning involves making an observation about something specific and applying it to the larger population. Although the premise may be correct, the conclusion may not be correct.
  - **Assumption/Observation 1:** My cats all have short hair.
  - **Conclusion:** All cats have short hair.

3 | **Developing and organizing the experiment.** There are several questions to consider. How will participants be selected? How many participants are needed? What will participants do? In clinical research, studies often involve a **control group** and an **experimental group.** The two groups are identical in composition with the exception of the aspect called the **independent variable**.

- In the experimental group the independent variable will change.
- In the control group the independent variable is constant and will not change.
- The **dependent variable** is the effect of manipulating the independent variable. It is what you measure in the experiment and may or may not change in response to the independent variable.

The control group and experimental group are compared against each other in an experiment. For example, when testing a new drug, the experimental group will be given the actual drug and the control group may be given a **placebo.** A placebo is a substitute used to determine if the effects one may show are due to the belief that they are being tested.

4 | **Data collection.** This is where the results of your observations and testing are recorded. The method of data collection must be identified, such as using a meter stick, scale, graph, or calculator. No matter the method, it must be consistent, controlled, and accurately recorded.

5 | **Data analysis.** The data that is collected must be analyzed to determine if the intervention is **statistically significant.** Statistical significance indicates whether the results of a study happened by chance or by some other factor. If you have statistical significance then you can be confident that the results were ***less*** likely to be caused by random chance. However, care must be taken to not extrapolate statistically significant data beyond the bounds of the trial until further studies among different populations can corroborate the findings. For example, you run an experiment where half of the subjects are asked to sit on the couch (control group) and the other half are asked to walk (experimental group) between 4 and 5 p.m. every day. You take the subjects' blood pressure (BP) at the start of the study and again at the end of the study and compare results between the two groups. The walking group has a statistically significant decrease in BP. This means you can be confident that it was the intervention (walking), not random chance, that lowered the BP.

6 | **Interpretation of results and drawing conclusions.** Using the data obtained from the experiment, one determines if the results support the original hypothesis and can logically explain how the data does so.

7 | **Accepted as theory.** If all criteria are met, the hypothesis may be accepted as a **theory.** This means that the hypothesis has been tested, is repeatable, unbiased, and has a well-substantiated explanation to the question or problem.

The scientific method is utilized throughout the world of medicine. As health care providers, an accurate diagnosis can be made through the observation and evaluation of a patient's signs and symptoms, which will then lead to a determination of appropriate medical care. The process of obtaining a correct diagnosis follows a sequence of steps that will include obtaining their medical history, performing a physical examination, and performing diagnostic tests. Based on the diagnosis, a treatment plan is formed.

Whatever treatment plan is developed, it should be based on scientific research and evidence. For example, medications used to treat a disease go through years of research and testing before determined to be safe and effective for the public. Clinical research is considered to be credible due to the rigorous application of the scientific method that has produced repeatable, proven results. Without it, we would not see advances or trends in medicine, nor could we understand the limits and challenges faced in clinical care.

Laboratory 1
Scientific Method Assignment
The Scientific Method

Name: ____

1 | Explain the difference between a hypothesis and a theory.

2 | Explain the difference between deductive and inductive reasoning.

3 | What is the difference between a control group and experimental group? What is the similarity?

4 | Describe what it means to have bias in a study.

5 | Explain the difference between a dependent and independent variable.

6 | Does the process of obtaining a correct diagnosis involve deductive or inductive reasoning? Explain your answer.

7 | Using the scenario provided, answer the questions that follow.

Observation: *My gray cat has diabetes. My gray cat is male and overweight.*

Hypothesis: *All male gray cats have diabetes.*

a | Is this hypothesis testable? Why or why not?

b | Do you expect that the results of an experiment would be repeatable? Why or why not?

c | Is this hypothesis biased? Explain your answer

8 | Using the scenario provided, answer the question that follows.

Observation: *Aliens have larger eyeballs than humans.*

Hypothesis: *Aliens have better vision than humans.*

a | Is this hypothesis testable? Why or why not?

9 | Using the scenario provided, answer the questions that follow.

A recent study performed in California on 100 local women discovered that there is a higher risk of lung cancer in women who use scented candles in their homes.

a | Is this study biased? Explain why or why not.

b | If one concludes that "scented candles cause lung cancer in women," is this conclusion based on inductive or deductive reasoning? Explain your answer.

c | What question(s) might you ask the person who conducted this study?

10 | Using the scenario provided, answer the questions that follow.

A pharmaceutical company is testing a new drug to treat hypercholesterolemia (excess cholesterol in the blood). The experiment involves 5,000 people who are over the age of 40 and have been diagnosed with hypercholesterolemia in the past year. All participants have a normal BMI, exercise 2–3× per week, are employed full-time, and do not have any other major underlying health conditions. The population profile includes both genders, is racially and ethnically diverse, and includes participants from five states in the mid-west United States. Half of the participants were given the new drug, the other half were given placebo, and both groups were monitored over the course of two years. All participants were required to eat a oatmeal 3× per week. No other dietary modifications were required. Results of the study show that those given the drug had an average of a 20% decrease in blood cholesterol levels, while those that took the placebo had a 5% decrease. Side effects of those who took the drug included joint pain, headaches, and stomach pain. Side effects of those taking the placebo included headaches.

a | Identify the independent and dependent variables in this experiment.

b | The placebo group demonstrated a slight decrease in blood cholesterol and experienced headaches. What may account for this?

c | Do you think there was bias in this study? Explain your answer.

d | What question(s) might you ask the person who conducted this study?

2 Measurements

Objectives

- Understand the terms ***precision*** and ***accuracy*** and why they are both important for making valid measurements.
- Take measurements and perform conversions using the metric units for length, mass, and volume.
- Perform conversions between Fahrenheit and Celsius temperature measurements.
- Become familiar with the scale of various structures and fluids composing the human body.

Materials Needed to Complete This Lab

- **Volume**
  - Graduated cylinders
  - Beakers
  - Transfer pipettes
  - Syringes
- **Mass**
  - Laboratory balance
- **Length**
  - Ruler
  - Meter stick
- **Temperature**
  - Thermometer
  - Water bath incubator
  - Ice water bath

Lab Activity 1 | Accuracy and Precision

Definitions

- **Accuracy** is the degree to which a measurement meets the true or accepted value.
- **Precision** is how close two or more measurements of the same item are to each other.

Procedure

Evaluate the targets below and determine whether they exhibit 1) high or low accuracy and 2) high or low precision.

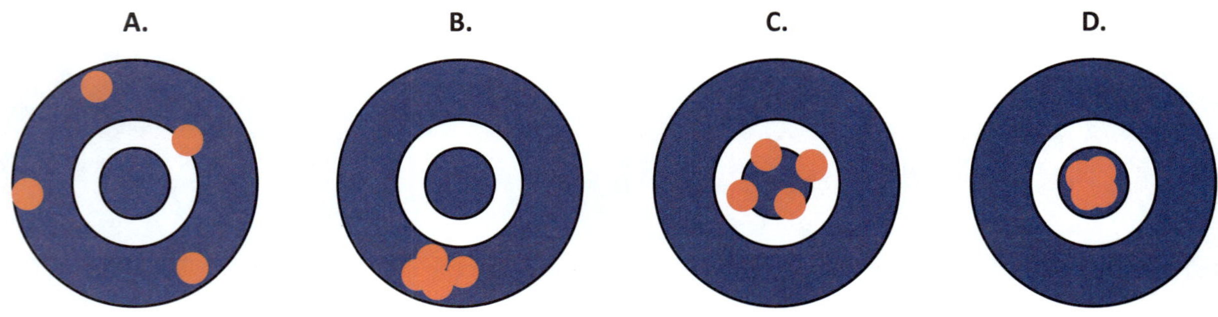

Figure A	Accuracy: High or Low	Precision: High or Low
Figure B	Accuracy: High or Low	Precision: High or Low
Figure C	Accuracy: High or Low	Precision: High or Low
Figure D	Accuracy: High or Low	Precision: High or Low

Lab Activity 2 | Conversions

In science, we use the metric system when taking measurements. This is a convenient system to use as it is based on units of ten. Later in this lab, you will be introduced to using the metric system for taking measurements. In this lab activity, you will learn how to make conversions from one metric unit to another (i.e., from meters to centimeters). The information in the following tables will help you as you get familiar with the prefixes and their multiple or division of the base metric units.

Table 2.1 | Multiples and Divisions of Metric Units

Multiplication Factor	Prefix	Symbol
$1{,}000{,}000{,}000 = 10^9$	giga	G
$1{,}000{,}000 = 10^6$	mega	M
$1{,}000 = 10^3$	kilo	k
$100 = 10^2$	hecto	h
$10 = 10^1$	deca	da
Base $1 = 10^0$	—	—
$0.1 = 10^{-1}$	deci	D
$0.01 = 10^{-2}$	centi	c
$0.001 = 10^{-3}$	milli	m
$0.000001 = 10^{-6}$	micro	μ
$0.000000001 = 10^{-9}$	nano	n

Table 2.2 | Base Metric Units

Base Quantity	Name	Symbol
Length	meter	m
Mass	kilogram	kg
Volume	liter	L

Temperature Conversion Formulae

°Celsius to °Fahrenheit: $°F = \frac{9}{5}°C + 32$

°Fahrenheit to °Celsius: $°C = \frac{5}{9}(°F - 32)$

When converting from centimeters (cm) to meters (m), you multiply by 0.01 (or 10^{-2}). If you are converting from micrometers (μm) to meters (m), you multiply by 0.000001 (or 10^{-6}). And, from kilometers (km) to meters (m), you multiply by 1000 (or 10^3).

For example, if you want to convert 54 centimeters (cm) to millimeters (mm), you can look at the table, and determine there are 10 mm in 1 cm.

$$54\ \cancel{\text{cm}} \times \frac{10\ \text{mm}}{1\ \cancel{\text{cm}}} = 540\ \text{mm}$$

Keep in mind the units that you are converting from cancel out, and you are left with the units into which you are converting for your answer. If your desired units are not on top, recheck your equation! Also, when converting from a larger unit to a smaller unit, the result will be a larger number. If you are converting from a smaller unit to a larger unit, the result will be a smaller number. Another way to think about it is which direction (left or right) you are moving the decimal point, and how many spaces you need to move it.

Lab Tip: You can make this simpler by using the chart to determine how to move the decimal point. For our example above converting 54 cm to mm, we are converting from a larger unit to a smaller unit, so our answer will be larger—we need to move the decimal point one space to the right since a centimeter is one unit of ten larger than a millimeter: **54 cm = 540 mm.**

If we are converting 54 cm to m, we are converting from a smaller unit to a larger unit, so our answer will be smaller—we need to move the decimal point two spaces to the left since a centimeter is two units smaller than a meter: **54 cm = 0.54 m.**

Practicing Metric Conversions

1 | 1 meter (m) = __________ centimeters (cm) = __________ millimeters (mm)

2 | 5.0 kilograms (kg) = __________ grams (g) = __________ milligrams (mg)

3 | 96 centimeters (cm) = __________ millimeters (mm) = __________ micrometers (µm)

4 | 200.0 microliters (µl) = __________ milliliters (ml) = __________ liters (L)

5 | 400 micrograms (µg) = __________ milligrams (mg) = __________ grams (g)

6 | 37°C = __________ °F

7 | 212°F = __________ °C

Lab Activity 3 | Taking Measurements Using the Metric System

For this part of the lab, you will go to the designated stations for measuring length, volume, mass, and temperature. Be sure to take your time so you take accurate measurements. You can compare your measurements with your classmates to see how precise the class is.

Caution—Lab Safety And Etiquette Note: Be sure to clean up after yourself whenever you are working in a lab. If you spill something, notify your instructor and work to clean it up. Be careful when working with glassware. If a piece of glassware breaks, notify your instructor.

A | Measuring Length—Basic Unit is the Meter (m)

For this activity, you will measure the length of different bones of the human skeleton. This will introduce you to some of the bones and structures you will encounter in future labs. You will use the meter sticks and rulers available at the station to take the following measurements. Record your results in the table.

Table 2.3 | Measurement Activity

Measurement Activity	Result
Length of femur (from greater trochanter to lateral condyle); record measurement in centimeters (cm)	
Width of proximal tibia; record measurement in centimeters (cm)	
Length of 1st metatarsal; record measurement in centimeters (cm)	
Length of foot's 5th distal phalanx; record measurement in millimeters (mm)	
Convert your femur measurement to meters (m)	
Convert your phalanx measurement to micrometers (µm)	

B | Measuring Mass—Basic Unit is the Kilogram (kg)

For this activity, you will measure the mass (amount of matter in a substance) using an electronic balance. Be sure the surface of the balance is free of debris and that you zero (tare) the balance prior to taking your measurement. Place your "specimen" on the balance and record the measurement in grams (g). Remember 1 kg = 1000 g. Record the mass and name of the three specimens of your choice from those available. "Specimens" being measured are representative of the average mass of those organs in the human body. Record your results in the table below.

Table 2.4 | Measuring Mass of a Specimen

Specimen Name	Mass (g)	Conversion
1.		
2.		
3.		

Extra: Convert one specimen's mass to kilograms and one to milligrams. Show your work below.

C | Measuring Volume—Basic Unit is the Liter (L)

For this activity, you will measure volumes of water using graduated cylinders and beakers. You will notice the graduated cylinder is marked with horizontal lines to denote its specific unit (i.e., ml or L). You will measure the volume up to the line indicating your desired volume. When doing this, you need to be aware that one property of water is adhesion—so it sticks to the sides of the graduated cylinder or beaker. Therefore, when measuring a volume (whether in the lab or in your kitchen), you look for the meniscus and take your reading from the bottom of the meniscus.

1 | Using the beaker provided, measure 0.5 L of water.

2 | Transfer 100 ml of the water to a graduated cylinder.

3 | Using a transfer pipette, transfer 10 ml from the first graduated cylinder to a second, smaller, graduated cylinder.

Think about which of these pieces of lab equipment is the most accurate to use for small volumes.

4 | Measurement of simulated body fluids:

a | **Measurement of the volume of cerebrospinal fluid (CSF):**

- Using a graduated cylinder, measure 150 ml of water and pour it into a beaker.
- This measurement represents the average volume of CSF circulating in the human body at any given time.

b | **Measurement of the volume of sweat:**

- Using a graduated cylinder, measure 200 ml of water and pour it into a separate beaker.
- This measurement represents the average volume of water lost as sweat per day.

c | **Measurement of urine volume:**

- Your instructor will pour 1500 ml of water into a beaker.
- This measurement represents the average volume of water lost as urine per day. (**Note:** This volume can vary considerably.)

d | **Considering the three volumes of fluid, answer the following questions:**

- Are these volumes of fluid what you would have anticipated finding in the human body? Why or why not?
- What could affect the volume of sweat?
- How would dehydration affect the volume of urine?

Very small volumes of a liquid are often measured using a syringe. This is a common method used to measure and administer liquid medications, and they are measured as milliliters (ml) or cubic centimeters (cc). 1 ml is equivalent to 1 cc.

5 | Using a plastic syringe, measure 2.5 cc of the liquid provided, and perform the following conversions:

a | 2.5 cc = __________ milliliters (ml)

b | 2.5 cc = __________ liters (L)

D | Measuring Temperature—Using Thermometer with the Celsius Scale

For this activity, you will measure the temperature of water in three different conditions. Once again, take your time to get the most accurate reading you can. Record your results in the table below.

Table 2.5 | Temperature of Water in Different Conditions

Measurement	Result °C
Beaker of water—room temperature	
Beaker of water—ice bath	
Beaker of water—warm water incubator	

3 The Microscope

Materials Needed to Complete This Lab

- Compound light microscope
- Prepared slide: newsprint "*e*"
- Prepared slide: colored cross-threads
- Clean, disposable microscope slides
- Disposable plastic coverslips
- Methylene blue stain
- Toothpicks
- Lens paper

A | Parts of the Microscope

1 | The microscope that is used in the lab is a ***compound light microscope.*** The lenses through which you will view the specimens are known as the ***oculars*** or ***eyepieces***. These lenses magnify the object you are viewing ten times (10×). You will note that the number 10× is written on the barrel of the oculars. Your microscope is considered to be ***binocular*** because it has two oculars (bi- means two). The binocular eyepieces can be adjusted to fit the width of your own eyes. When looking through the eyepieces after they have been properly adjusted for your eyes, you should only view one "circle" of light (or field of view). If you see two circles of light or overlapping circles, you will need to make further adjustments to the width of the eyepieces until only one circle is visible.

2 | The oculars are connected to a box-like structure called the ***binocular tube***. At the bottom of this structure is a ***revolving nosepiece*** that has three magnifying lenses attached. When moving from lens to lens, always rotate the nosepiece by grasping the disc which rotates; ***never*** grab the lenses attached to the disc.

3 | Each of the magnifying lenses that are attached to the revolving nosepiece is known as an ***objective lens***. Each objective lens has its own ***magnification power*** which is marked on its barrel. The objective lenses on your microscope include the following:

- **4×—Scanning objective:** This lens has the shortest barrel and is used to initially focus the slide or to locate a certain area of the slide. Always begin your study with this lens.

- **10×—Low power objective:** This lens is used to more closely view the object of interest after initially focusing with the scanning objective.
- **40×—High power objective:** This objective is used for very close focusing.

When viewing an object, it is magnified in two locations: the oculars magnify the specimen 10× and each of the objective lenses has additional magnification power. Therefore, the ***total magnification*** of the specimen that you view can be determined by multiplying the ocular power by the magnifying power of the objective lens in use.

For example: ocular magnification = 10× and the high power objective magnification = 40×. The total magnification = 10× × 40× = 400×.

4 | All slides you examine will be placed upon the horizontal ***stage*** of the microscope. The hole in the stage allows light to pass through the slide. A slide will be held in place on the stage by ***specimen holders.*** You will move the slide on the stage by rotating the ***stage adjustment knobs*** which extend perpendicularly from the right side of the stage. One knob moves the slide from side-to-side while the other knob moves the slide up-and-down.

5 | In order to best view a specimen, a light source is required. The ***lamp*** of the microscope provides the light which projects up through the stage. The lamp is located in the base of the microscope. On the side of the microscope's base, there is a ***light regulator control dial;*** this dial adjusts the amount of light emitted by the lamp. Turning the dial towards the front of the microscope makes the light brighter, while turning towards the back of the microscope makes it dimmer. The dial should be set about midway. The light from the lamp passes through a ***condenser,*** which intensifies or concentrates the light source before it passes through the specimen.

Optimal lighting is essential for precise viewing of microscopic specimens. Sometimes very bright light works best, and other times dim light works best. While using your microscope, try adjusting the brightness of the light by the following methods:

- Adjust the light from the lamp by turning the light intensity control dial up or down.
- The condenser contains an ***aperture iris diaphragm lever*** that moves from side-to-side. By moving this lever, the amount of light passing through the condenser is adjusted. This is similar to adjusting the light passing through the pupils of your eyes.

B | Care and Handling of the Microscope

The ***arm*** of the microscope is where the stage is connected. To carry the microscope, grasp the arm with one hand and place your other hand beneath the base. ***Never*** carry the microscope with one hand only.

Because many students use the microscopes, you might find your microscope lenses in need of a light cleaning. Using the provided ***lens paper,*** you can gently wipe the oculars, the bottom of each objective lens, and the microscope slides. If these are extremely dirty, you may dampen, but not saturate, your lens paper with a bit of water. In some cases, it may require additional cleaning by the laboratory technician. ***Never*** use tissues, paper towels, or material other than the provided lens paper to clean the lenses, as these materials will scratch the glass.

C | Using the Microscope

1 | **Focusing:** Located on each side of the arm of the microscope, immediately above the base, you will find a set of two knobs—large and small. The large knobs closest to the arm are the ***coarse adjustment knobs.*** The small knobs attached to the ends of the coarse adjustment knobs are the ***fine adjustment knobs.*** These are used to focus the specimen by moving the stage of the microscope closer to or farther away from the objective lens.

Always begin focusing specimens with the scanning objective in place (facing the stage) and the stage at its lowest position. Move the stage to its lowest position by turning the coarse adjustment knobs. (If the stage does not move down easily, the microscope may be locked; consult your instructor.)

Lab Activity

1 | Letter "*e*" Slide—Learning the Basics

a | Plug in and turn on the microscope using the ***power switch*** located on the back of the base. Once the power switch is turned on, you should see light from the lamp. Occasionally, the lamp may be on but the light too dim to detect. Try rotating the light adjustment dial on the left side of the microscope base by turning the dial approximately half way. If the lamp does not turn on, try another outlet. If there is still no light present, the lamp may be burned out; consult your instructor.

b | Prior to placing the slide on the microscope, move the stage to its lowest position by using the coarse adjustment knobs and rotate the revolving nosepiece so that the scanning objective faces the stage. (You should hear and feel a "click" when the objective is in the proper position.)

You will always begin an examination of each slide with the stage in its lowest possible position and the lowest objective (scanning objective) in place!

c | Pick up the prepared slide of the letter "*e*" and hold it by the edges. Determine if it contains smudges or fingerprints, and, if so, clean it with the lens paper as described previously. When clean, position the slide on the stage with the specimen holder keeping it firmly in position. Using the stage adjustment knobs, center the slide on the stage over the opening through which the light passes.

d | Adjust the oculars so that they are the appropriate width for your eyes. Keeping your eyes approximately one inch away from the oculars, look through them. You should see one circular field of vision. You should also notice a "pointer" in one of the oculars. This thin, straight wire is typically located in the right ocular (because most people are right-eyed dominant) and is a valuable tool for use in drawing attention to a particular area of the specimen. If you see more than one circle of light, see shadows, or do not see the pointer, then you need to either make further adjustments of the ocular width or move your eyes closer to or farther away from the oculars. If you still do not see the pointer, you may be left-eyed dominant. Consult your instructor about switching the position of the oculars. Try to keep ***both eyes open*** when viewing the slide.

e | Using the coarse adjustment knobs, slowly raise the stage until the letter "*e*" comes clearly into focus. You will have to raise the stage quite a distance. Fine-tune the focusing with the fine adjustment knobs. Once the "*e*" is in focus, rotate the revolving nosepiece to the low power objective (10×). ***Do not use the coarse adjustment knobs at this point.*** Using the fine adjustment knobs only, bring the image back into precise focus.

f | You may notice that the letter "*e*" is not in the same orientation as you originally positioned the slide on the stage. The letter "*e*" is upside down. This is due to the design of the lenses which will turn the specimen upside down and cause it to be mirror reversed. Keep this in mind as you view the slides. This also explains why, as you move the slide with the stage adjustment knobs, the specimen appears to move in the opposite direction.

g | Once precisely focused, rotate the nosepiece to the high power objective (40×) and focus using the ***fine focus knob only.*** Try adjusting the light sources as described previously and notice how the specimen changes. Your examination of the letter "*e*" slide is now complete.

h | To remove the slide from the stage, first rotate the revolving nosepiece to the scanning objective (4×). Then lower the stage to its lowest position using the coarse adjustment knobs. Now you may remove the slide from the stage. **Failure to perform these steps can result in cracked and broken slides!**

Reminder: Never use the coarse adjustment knobs with any objective except the lowest power (4×).

The microscope lenses are considered to be ***parfocal*** which means they have been designed so that once the specimen is in focus with the scanning objective in place (4×), additional focusing on higher objectives will require only fine tuning with the fine adjustment knobs.

Lab Tip: If you "lose your specimen" under the high power objective, return to the scanning objective (4×), re-center the slide, and begin the step-by-step focusing process again.

2 | **Crossed Threads:** The next slide you will examine contains three colored crossed threads. Following the same procedure as you used in the examination of the letter "*e*" slide, initially focus the slide. Using the fine focusing knobs, slowly move it up and down slightly in order to determine the different depths of the threads. Determine which thread is on the top, middle, and bottom.

3 | **Wet Mount Preparation**

a | The final procedure consists of the collection and preparation of a slide of your own cells.

b | Using a clean toothpick, gently scrape the inside of your cheek. Smear the scrapings on a clean plastic disposable slide. Place a drop of methylene blue stain directly on top of the cheek scrapings. (Be careful not to get the stain on your clothing.)

c | Using a square plastic cover slip, angle one edge of the cover slip at the edge of the methylene blue stain on the slide. ***Gently*** allow the cover slip to drop into place over the stain and the cheek scrapings. Placing the cover slip on the slide in this manner will help to eliminate air bubbles from being trapped beneath the cover slip.

d | Examine the specimen under the microscope using the same technique you used with the prepared slides.

e | You should see cells that look like reptile scales or "fried eggs." These are called squamous epithelial cells and form the lining of mucous membranes in the body (such as in your mouth).

f | The cells tend to clump together and sometimes fold over on themselves. The dark circle in the center of each cell is the nucleus. You may see small black specks or thread-like material in your slide. This is debris from your mouth or materials on the microscope slide. Keep looking for your cheek cells!

g | When finished viewing this slide, consult your lab instructor on disposal.

D | Labeling of Slides

You will be viewing many slides during this course. Prior to examining the slides, note the information given on the slide. The information includes the type of material mounted on the slide and the type of "cut" made of the specimen prior to mounting. It may say "whole mount" which indicates the entire specimen is on the slide. Or you may see the following terms indicating the manner in which the specimen was prepared: **c.s.** means

cross section of the specimen (such as cutting across the top of a dry spaghetti noodle); **l.s.** means ***longitudinal section*** (as if you were cutting the length of the spaghetti noodle); **sec**. indicates ***section*** in which you may see several views (cross sections and longitudinal sections) on the same slide.

E | Procedure for Storing the Microscope Correctly

To prevent damage to the microscope, please follow the procedures below for returning the microscope to the lab cabinet.

1 | Position the scanning objective (4×) so that it faces the stage.

2 | Use the coarse adjustment knobs and move the stage to its lowest position.

3 | Turn the lamp to the lowest setting and turn off the microscope.

4 | Put dust cover on the microscope.

5 | Using two hands, place the microscope in the lab cabinet with the ***arm facing out.***

Summary for Correct Procedure for Getting Out and Using the Microscope:

- Always carry the microscope with **two hands**—one supporting the **base** and one holding the **arm.**
- Once the slide is in position, **always start focusing with the 4× scanning objective** (the shortest "tube").
- **Only use lens paper** to clean an ocular, an objective lens, or a slide.
- Only use the **coarse adjustment knobs** with the **4× objective.**
- When moving the microscope on the lab table, **pick it up.** Sliding the microscope across the table top can cause slides to break and damage to the lamp.

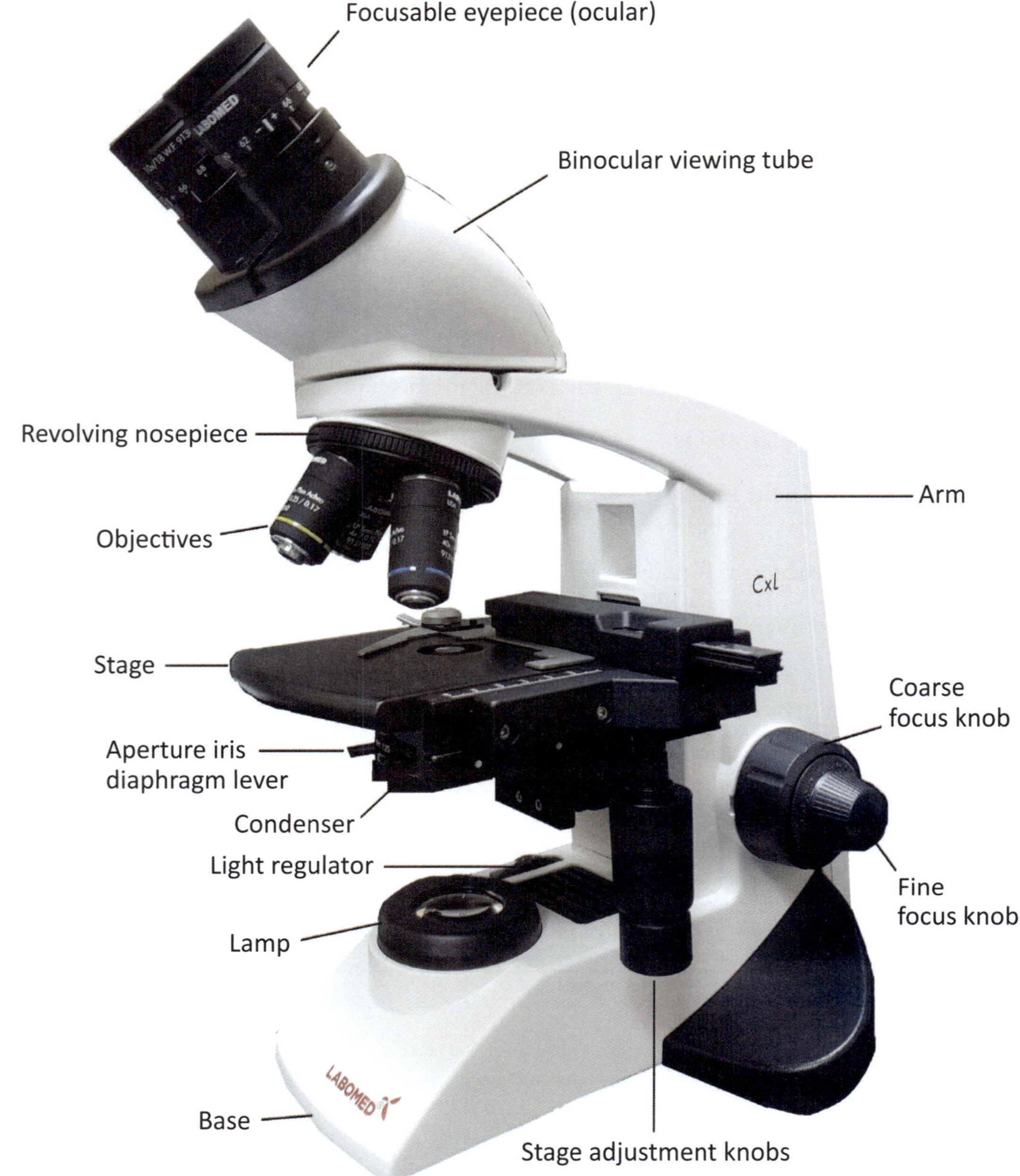

Figure 3.1 | Parts of the Microscope

Laboratory 3 Review Questions

The Microscope

Name: ____________________

1 | Match the following:

________ condenser

________ iris diaphragm

________ fine adjustment

________ ocular

________ objective

a | a magnifying lens found on the revolving nosepiece

b | eyepiece

c | used for refinement of detail in focusing an image

d | a structure that allows for varying amounts of light to pass through the slide

e | a system concentrating light from an illumination source

2 | Fill in the appropriate magnification power in each of the blanks.

Magnifying Power of Ocular	Magnifying Power of Objective Lenses	Total Magnification
	Objective 1 (scanning)	
	Objective 2 (low power)	
	Objective 3 (high power)	

3 | Which microscope objective gives the widest field of view (shows most of the object being viewed)? Which gives the smallest field of view (shows the least amount of the object being viewed)?

a | Widest field of view: ____________________

b | Smallest field of view: ____________________

4 | Why should only lens paper be used to clean the oculars and the objectives?

5 | What is meant by the term ***parfocal?***

Multiple Choice. Circle the correct answer. (One answer per question.)

6 | A microscope ___________ the image of a specimen.

- *a* | magnifies
- *b* | reverses
- *c* | reduces
- *d* | erases
- *e* | both a and b are correct

7 | The part of the microscope that controls the amount of light penetrating the specimen is the:

- *a* | objective
- *b* | iris diaphragm
- *c* | eyepiece
- *d* | nosepiece
- *e* | specimen holder

8 | Initial focusing of any slide is done with the:

- *a* | low power objective
- *b* | high power objective
- *c* | scanning objective
- *d* | fine adjustment knobs

9 | Which of the following represents poor technique when using a microscope?

- *a* | storing the microscope with the scanning objective toward the stage
- *b* | turning the coarse adjustment knobs with the objective on high power while looking through the oculars
- *c* | looking through the oculars with both eyes open
- *d* | wiping the external surfaces of the oculars and objectives with lens paper prior to using the microscope

10 | While viewing a slide through the microscope, moving the slide to the left makes the object appear to

- *a* | move right
- *b* | move left
- *c* | remain stationary

4 pH—Acids, Bases, and Buffers

Introduction

Understanding the concept of acids/bases and the quantification of them through pH is very important in physiology and in clinical situations. An alteration of the normal acid/base balance and the resulting disruption of the normal pH of a body fluid can have drastic consequences to normal cellular, histological, and organ functions. These functions are affected because a change in pH can negatively influence enzyme activity and thus the biochemical pathways in which these enzymes work. So, it is imperative that the pH of a fluid remain in check for normal function and homeostasis.

The pH of fluid is determined by the hydrogen ion concentration within that fluid. Recall that **acids** are substances that when placed in water, will **ionize** and release **H^+** into the fluid. The hydrogen ions will then begin to decrease the pH of the fluid; the fluid becomes acidic. This can be seen on the pH scale by a number lower than 7.

Conversely, a **base** is a substance that when put in water will release **hydroxyl ions (OH^-)** into the fluid. These hydroxyl ions will then cause the pH of the fluid to rise; the fluid becomes **basic** or **alkaline.** This can be seen on the pH scale by a number greater than 7.

The strength of an acid or base depends on the degree of ionization. **Strong acids or bases ionize completely** (release all of the H^+ or OH^- into solution). Weak acids or bases ionize partially or to a lesser extent than strong ones. In practical terms, the more H^+ ions in the solution the more acidic the solution is, and the more OH^- ions in solution the more alkaline the solution is.

Under normal homeostatic control, pH is very tightly regulated. The amount of hydrogen ions are kept at a very low level in the body fluids when compared to other ions. For example, in blood, sodium is around 142 mEq/L where a hydrogen ion is .00004 mEq/L. Also, the amount of fluctuation of H^+ that is tolerated in the extracellular fluid is very low when compared to sodium.

To resist the pH change, the H^+ and the OH^- must be removed or bound to another molecule to decrease their ability to change the existing pH. The human body accomplishes this by using a number of methods. The body can use the respiratory and renal systems to remove the ions. These are called **physiological buffers.**

Another method that is used, and is the primary method used by the body fluids directly, is the use of **chemical buffers.** A chemical buffer is a substance that counters pH changes by chemically reacting with the H^+ or OH^- ions, thus preventing them from changing the existing pH.

Different fluids of the body have a normal pH for that fluid. For example, blood has a pH range of 7.35–7.45 and urine has a pH range of 4–8. Stomach fluid can have a pH range of 0.8–2. The pH of these fluids would be considered normal as long as it remains within that range. If a pH shifts above or below the limits of the range, the fluid will not be able to work effectively. Thus, we need to keep the pH within these limits.

Buffer systems are systems that use a combination of chemicals to stabilize the pH of a fluid. Usually, the buffer system is a combination of two chemicals, a weak acid and salt of the weak acid. The salt can act as a base in the system. So, a buffer system is a combination of the weak acid and a weak base. From a functional aspect, the components of the buffer reversibly bind the free H^+ ions or the free OH^- ions in body fluid.

The general form of the buffering action is as follows: Buffer + $H^+ \leftrightarrow H^-$ Buffer or Buffer + $OH^- \leftrightarrow OH^-$ Buffer.

The weak base will attempt to bind and neutralize acids or H^+ ions that are floating freely in the fluid, and the weak acid will attempt to bind and neutralize the OH^- ions floating freely in the fluid. **It is this binding action that stabilizes or buffers the pH of the fluid.** The binding action removes the H^+ or OH^- ion from the solution and prevents the pH from dropping or rising respectively.

The buffering or neutralization action can be seen in the following illustration:

We will mix an acid, HCl, (this strong acid has a pH of 2) with a base, NaOH (this strong base has a pH of 14). The mixing results in the following reaction:

$$HCl + NaOH \rightarrow H_2O + NaCl$$

Note that the reaction led to the formation of water (H_2O) and table salt (NaCl). Both of these products have a pH of 7. Thus, the reaction resulted in the neutralization of both the original acid and base! This reaction shows how an acid or a base can be used to neutralize another base or acid respectively.

A very useful tool that allows one to see how much the pH of a fluid changes or how well a buffer system works is the following equation:

$$pH = pK_a + \log \frac{[base-]}{[acid+]}$$

By interpreting the equation, one can predict the potential change in pH. By knowing how much acid or base is added to the fluid, one can find the resulting change and determine how much the pH changed. This equation can also be used to determine how well a solution with a buffer can resist the pH change when an acid or a base is added to the fluid.

Before proceeding, answer the following questions.

1 | Using the equation, predict how the pH will change by the addition of an acid to a solution and the addition of a base to a solution.

2 | Explain how water can behave as an acid and/or a base.

The purpose of this lab is to gain an understanding of acids and bases, and how they are buffered in body fluids.

Materials Needed to Complete This Lab

- 150 ml beakers (5)
- pH meter
- Distilled water
- Hydrochloric acid (HCl)
- Sodium hydroxide (NaOH)
- Standard buffer solution (pH 7)

Lab Activity

Caution: Be extremely careful when handling the concentrated HCl and NaOH to avoid injury!

1 | With a wax pencil, label the beakers 1–5.

2 | Fill beakers 1, 2, and 3 with 75 ml of distilled water.

3 | Fill beakers 4 and 5 with 75 ml of standard buffer solution.

4 | In beaker 2, place 1 drop of HCl into water.

5 | In beaker 3, place 1 drop of NaOH into the water.

6 | In beaker 4, place 1 drop of HCl into the buffer solution.

7 | In beaker 5, place 1 drop of NaOH into the buffer solution.

8 | You will use the pH meter to measure the pH of each solution. ***It is very important to rinse the pH sensor with deionized water prior to and after each measurement.***

9 | Now, using the pH meter that has been standardized to pH 7, determine the pH of each of the solutions in the beakers by placing the pH sensor (probe) into the solution. ***Remember to rinse the sensor with deionized water prior to and after each measurement.***

10 | Record the values in Table 4.1.

11 | Predict what will happen to the pH of the solutions in beakers 4 and 5 when additional HCl and NaOH are added, respectively. Record your predictions.

12 | In beaker 4, add 3 additional drops of HCl to the solution; record the pH in Table 4.1.

13 | In beaker 5, add 3 additional drops of NaOH to the solution; record the pH in Table 4.1.

a | Did the changes match your predictions?

b | How well did the buffer solution do in resisting the pH change when additional HCl and NaOH are added to the solution?

Table 4.1 | pH of the Solutions

Beaker Number	pH
1	
2	
3	
4	
5	
4 with additional HCl	
5 with additional NaOH	

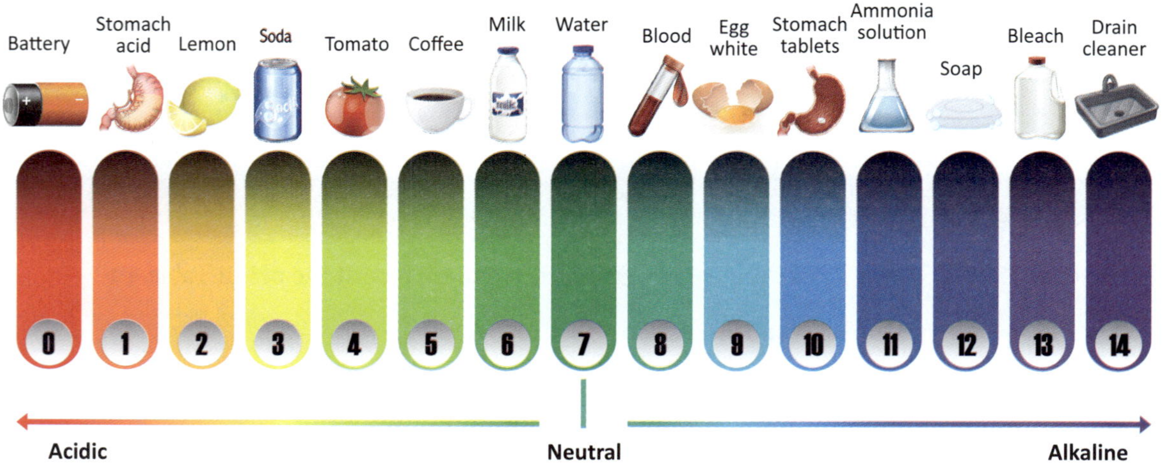

Figure 4.1 | The pH Scale

5

Membrane Transport

In order for organisms to survive, they have to carry out specific life processes. One of these processes is **movement.** Movement exists in many forms. For example, walking, running and breathing are forms of movement. But movement must occur at the molecular and cellular levels as well. Although not as readily apparent as walking and running, the movement at the molecular and cellular levels is vitally important.

The molecular or chemical movement is called **membrane transport** (also known as **biotransport**). Through membrane transport, chemicals are able to translocate through selectively permeable membranes.

There are two types of membrane transport—**active transport** and **passive transport**.

All ***active transport*** processes have two characteristics in common: 1) the molecules move from an area of low concentration to an area of high concentration; and 2) they require energy from the cell in the form of adenosine triphosphate (ATP).

All ***passive transport*** processes share two characteristics as well, but they are different from active processes: 1) the molecules move from an area of high concentration to an area of low concentration; and 2) they do not require the cell to expend energy (ATP).

Examples of passive transport include diffusion, facilitated diffusion, osmosis, and filtration.

This lab will focus on passive transport, specifically diffusion.

Passive Transport

Passive transport processes do require energy for them to take place, but the origin of the energy is not from the cell in the form of ATP. The energy for the movement originates in the molecules themselves. In other words, the molecules that are moving are providing the energy.

The energy for the movement is **kinetic energy.** All molecules contain this energy and are in constant motion. This inherent motion causes neighboring molecules in a region to collide with each other. These collisions cause the molecules to spread out and increase their distance from one another, much like the balls on a pool table when they are impacted. This molecular movement causing the collisions is called **Brownian movement.** Brownian movement is the foundation for all passive transport processes.

Anything that affects Brownian movement will have an effect on passive transport processes.

Before you proceed, answer the following questions:

1 | Define the term temperature.

2 | Predict how temperature might influence Brownian movement.

Diffusion is a passive transport process that applies Brownian movement. In addition to using the kinetic energy from Brownian movement, diffusion illustrates the other characteristics of passive movement—it involves a **concentration gradient.** If molecules of a substance are not evenly distributed in an area, this creates a situation where one region has more molecules and another region has fewer molecules. This uneven distribution creates a concentration gradient.

Molecules will follow this gradient and move from an area of higher concentration to the area of lower concentration until both regions have an equal distribution of molecules. When that occurs, **equilibrium** has been reached.

Diffusion is defined as the movement of molecules from a region of high concentration to a region of low concentration. The molecules follow or move down a concentration gradient. This will take place until equilibrium has been reached, meaning the number of molecules in both regions will be equal. In the human body the regions are often separated by a selectively permeable membrane that the molecules must pass through.

There are a variety of factors that can affect the movement of molecules and thus influence the rate of diffusion. By looking at the equation representing Fick's Law of Diffusion, one can get an appreciation for the variables and predict how changing the variables can affect the rate of diffusion.

Fick's Law of Diffusion: $\text{Rate} = K \times A \times \frac{(C2 - C1)}{D}$

K = Diffusion constant, based upon solubility and temperature

A = Surface area of membrane the molecules must move through (important in organs like the lungs, intestines, and nephron)

C2 – C1 = Concentration gradient

D = The thickness of the membrane the molecules must move through

Using the formula above, predict what a change in temperature, surface area, concentration gradients, or thickness of the membrane would have on the rate of diffusion.

- Temperature change: ____________________
- Surface area change: ____________________
- Concentration gradient changes: ____________________
- Thickness of membrane:: ____________________

Materials Needed to Complete This Lab

- 250 ml beakers (3)
- Food coloring
- Tap water (room temperature)
- Tap water (refrigerated)
- Tap water (hot)
- Timer

Lab Activity

1 | Using a wax pencil, label the beakers A, B, and C.

2 | In beaker A, fill three quarters full with refrigerated water.

3 | In beaker B, fill three quarters full with room temperature water.

4 | In beaker C, fill three quarters full with hot water **(be very careful to avoid getting burned).**

5 | Place the beakers on a white surface in order to better monitor the results.

6 | In beaker A, carefully place 1–2 drops of food coloring. Immediately begin timing.

7 | As equilibrium is reached in beaker A, record the time (in seconds) in Table 5.1.

- You will know that equilibrium has been achieved when the coloring is evenly distributed throughout the beaker.

8 | Repeat Steps 5 and 6 for beakers B and C while waiting for beaker A to reach equilibrium.

9 | Compare the time recorded for each beaker.

Table 5.1 | Time to Reach Equilibrium

Beaker	Time to Reach Equilibrium (seconds)
A (refrigerated water)	
B (room temperature water)	
C (hot water)	

1 | Did your results match your predictions?

2 | Explain your results.

6

Histology: The Study of Tissues

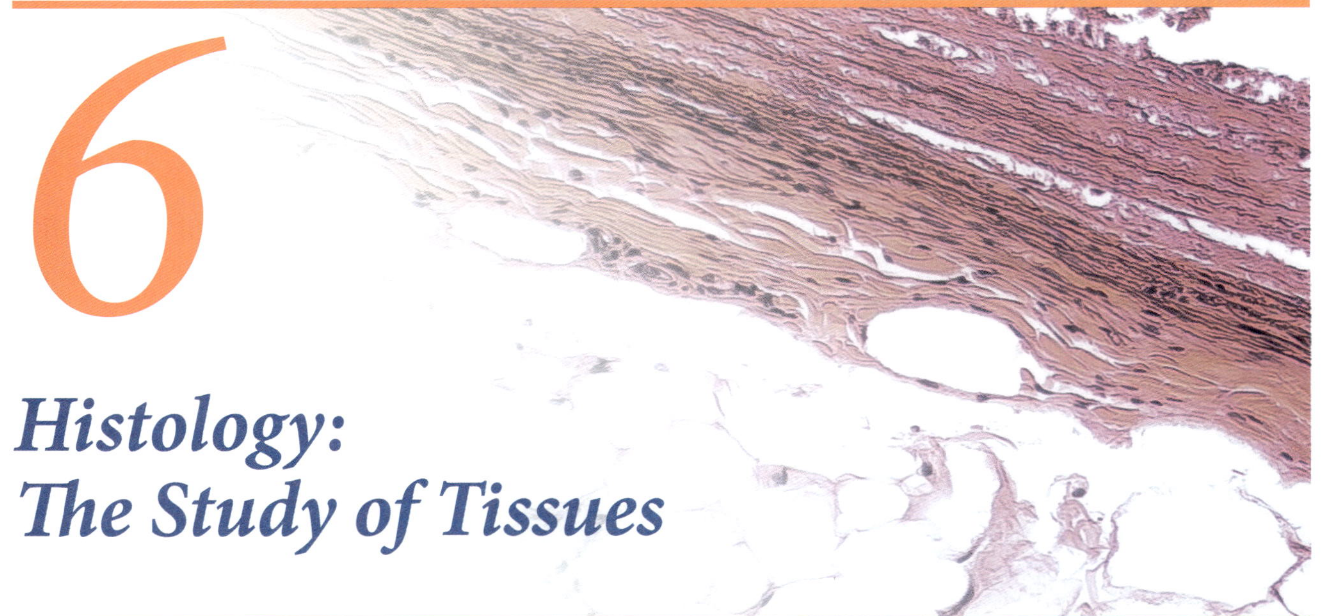

Tissues are collections of cells. These groups of cells organize and perform specific functions in the body. In humans, there are four categories of tissues: **epithelial, connective, muscle, and nervous.** The tissues are categorized based upon their appearance and functions.

> **Lab Tip:** When naming a tissue, you **must include its entire name,** including one of the four categories mentioned above. Examples: simple squamous *epithelial tissue;* hyaline cartilage *connective tissue,* cardiac *muscle tissue.* As you study a tissue, make a habit of repeating the entire name of the tissue.

Epithelial Tissue (Five Types to Identify)

Epithelial tissues are found covering the body's surface (e.g., on the skin), lining internal organs, and lining body cavities. Epithelial tissues also form membranes. Because of their locations, epithelial tissues are specialized in protection, secretion, absorption, and surface transport of materials.

General Characteristics of Epithelial Tissues

- The cells in epithelial tissue are packed tightly together, having little or no space between the cells (intercellular space).
- The most superficial cells in epithelial tissue have a free (or ***apical***) surface (faces an open space) at the top. The bottom-most basal cells are attached to a basement membrane.
- Epithelial tissues are ***avascular*** (contain no blood vessels). They must obtain their nutrients from the blood vessels found in underlying connective tissues.
- Because of their locations, epithelial cells are subject to considerable damage. So these cells are constantly being replaced by mitosis (cell division). In fact, they have some of the fastest rates of cell division in the body.
- Epithelial tissue is classified and named according to the shape of the cells composing the tissue and the number of layers in the tissue.

A | Shape of Cells

1 | ***Squamous:*** thin, flat, scale-like cells.

2 | ***Cuboidal:*** cube-shaped (appear nearly square or with rounded corners); the nucleus takes up most of the cell and is typically rounded.

3 | ***Columnar:*** tall and thin (like columns or rectangles standing on end); may be ciliated. The nuclei of these cells tend to be more oval.

B | Number of Layers of Cells

1 | ***Simple:*** 1 layer of cells.

2 | ***Stratified:*** more than one (or several) layers of cells.

3 | ***Pseudostratified:*** appears to have several layers, but is actually only one cell layer with cells of differing heights.

Lab Activity

Obtain slides for each of the following types of epithelial tissues. Using your textbook and other reference materials, be able to identify each of the different epithelial tissues. If you choose, you may make your own illustrations in the boxes provided.

Lab Tip: Do not use color as your only identification tool. Slides of the same tissue may have used alternate stains resulting in different colors for the same tissue type!

Types of Epithelial Tissues

1 | **Simple squamous epithelial tissue**

- **Description:** One layer of flat, tile-floor like cells. The nucleus tends to be centrally located in each cell.
- **Function:** Because this type of tissue is very thin, materials diffuse easily through it. It is also important in filtration.
- **Location:** Some locations for this tissue include the lungs (for movement of the gases oxygen and carbon dioxide) and in the body's capillaries.

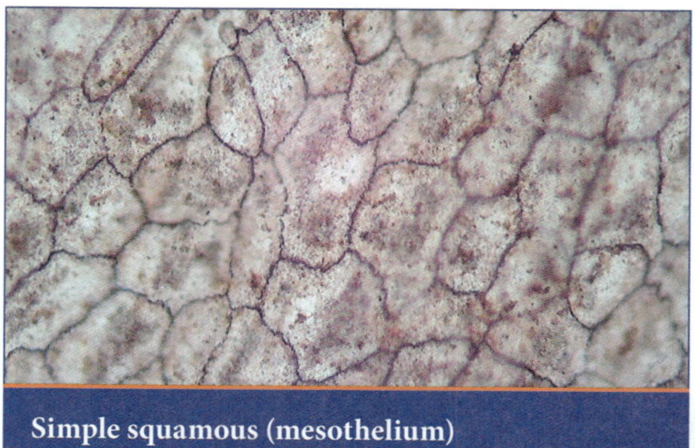

Simple squamous (mesothelium)

Simple squamous (mesothelium) drawing

2 | Stratified squamous epithelial tissue

- **Description:** Many layers of squamous cells. The cells look more rounded or cuboidal in the deep layers and flattened towards the superficial layers.

  The superficial layers of these cells are continuously being scraped or sloughed off (such as in the skin and in the digestive system organs).

- **Function:** The many layers provide protection.

- **Location:** Examples of this tissue are found in the outer part of the skin, the vagina, and both ends of the digestive system.

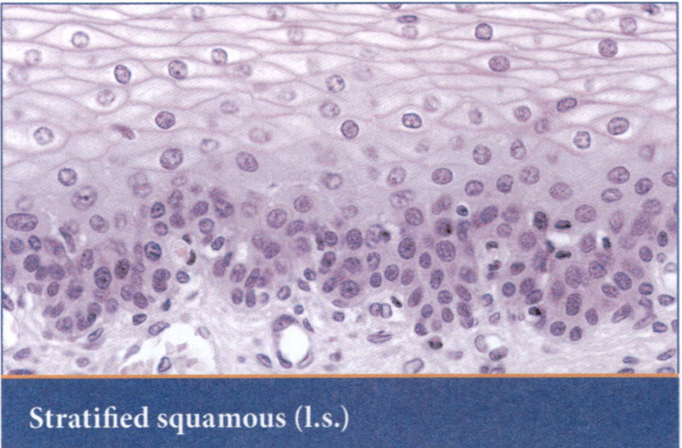

Stratified squamous (l.s.)

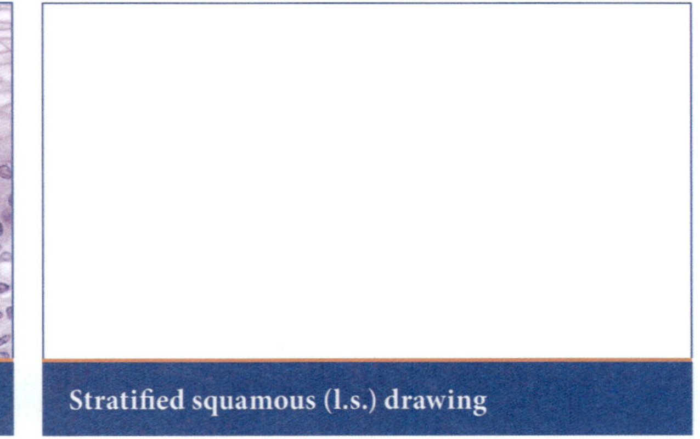

Stratified squamous (l.s.) drawing

3 | Simple cuboidal epithelial tissue

- **Description:** One layer of square cells. These cells often look more like squares whose corners have been rounded. Each cell has a fairly large, centrally located, rounded nucleus.

- **Function:** This tissue lines the ducts of glands which produce many of the body's secretions such as saliva and tears.

- **Location:** Simple cuboidal is found in many glands such as the tear glands, salivary glands, and sweat glands. It is also located in the small tubules in the kidneys.

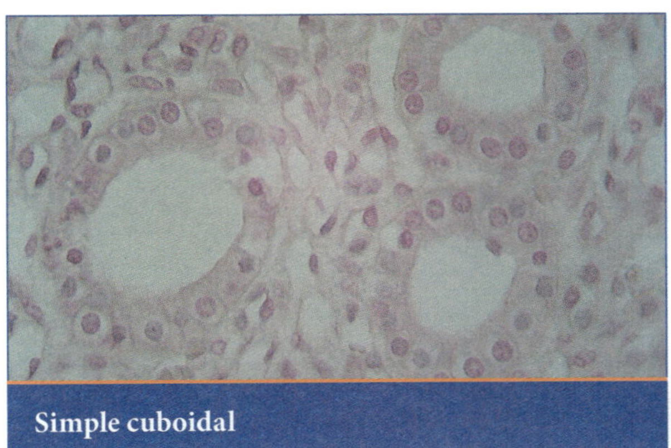

Simple cuboidal

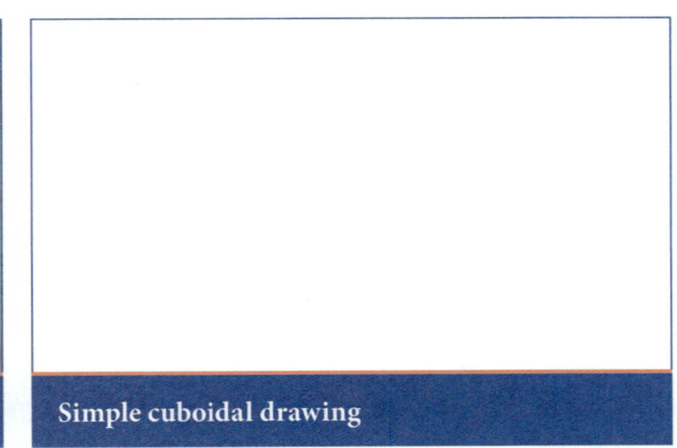

Simple cuboidal drawing

4 | Simple columnar epithelial tissue (ciliated & nonciliated)

- **Description:** One layer of columnar cells. The nuclei tend to occur in relatively straight rows through the tissues. Look for the oval-shaped nuclei near the bottom of each cell.

  May be ciliated or nonciliated. Look for the ***cilia*** (tiny hair-like extensions) on the apical surface of the cells.

  Simple columnar tissue often contains modified cells known as ***goblet cells*** (look like wine/water goblets) whose function is to secrete mucus.

- **Function:** Dependent upon location, this tissue may function in protection, absorption, lubrication, and movement of substances.
- **Location:** Parts of the digestive system (stomach, intestines), fallopian tubes, and respiratory system.

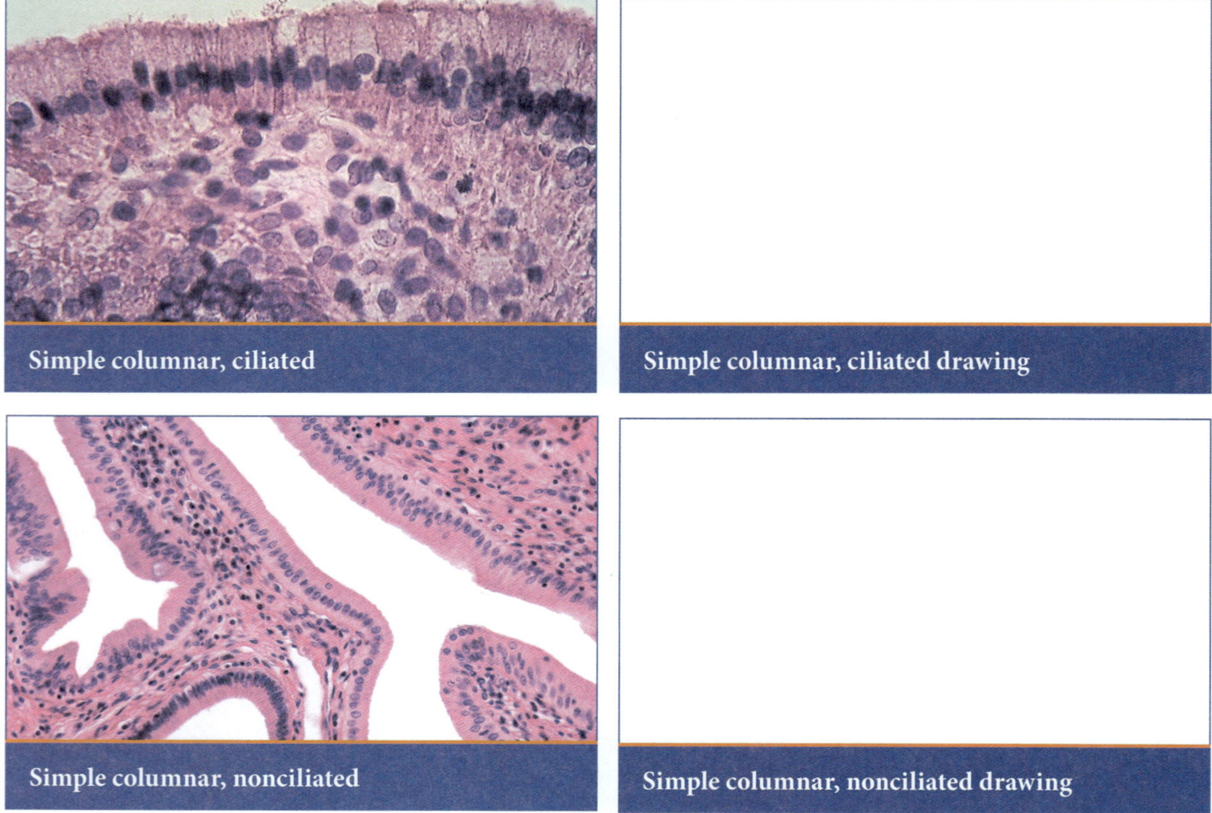

Simple columnar, ciliated

Simple columnar, ciliated drawing

Simple columnar, nonciliated

Simple columnar, nonciliated drawing

5 | **Ciliated pseudostratified columnar epithelial tissue**

- **Desciption:** One layer of columnar cells of differing heights. The nuclei in this tissue are scattered (irregular) rather than appearing in relatively even rows. Contains goblet cells.
- **Function:** Secretes mucus to trap debris and cilia to carry away the mucus and debris.
- **Location:** This tissue can be found lining many of the respiratory system passages.

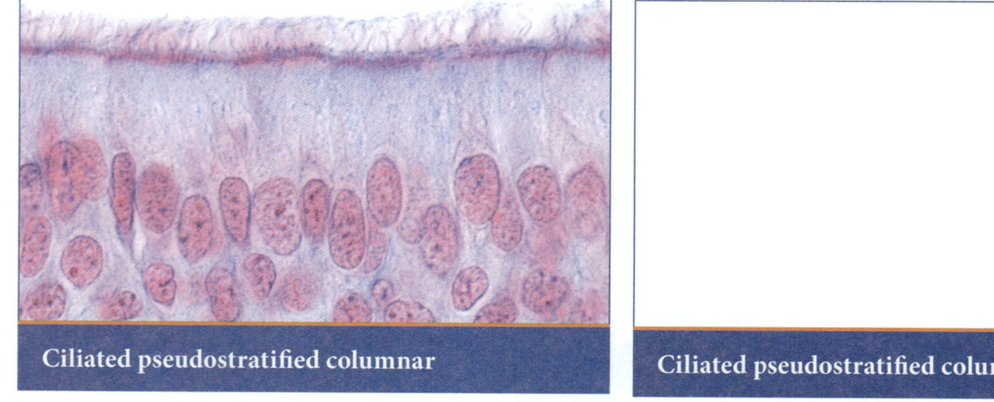

Ciliated pseudostratified columnar

Ciliated pseudostratified columnar drawing

Lab Tip: Don't forget to practice saying and writing the **entire** name of the tissue when you view the slides.

Connective Tissue (Eight Types to Identify)

As the name implies, connective tissue is often found connecting or holding together other tissue types or body parts. Connective tissue is extremely widespread throughout the body and is very diverse in its structure and functions.

General Characteristics of Connective Tissues

- In contrast to epithelial tissue which has tightly packed cells, connective tissue contains relatively few cells which are ***not*** densely or tightly joined together.
- Connective tissue cells tend to be widely separated by the intercellular substance known as ***matrix***.
- The matrix is a nonliving chemical mixture (ground substance) which varies dependent upon the type of connective tissue. The matrix may also contain a variety of nonliving protein fibers.
- Connective tissue can have a blood supply.
- Connective tissue does not cover surfaces or line organs so the cells do not have an apical surface. Some connective tissues are found sandwiched between other tissue types.
- Connective tissue is classified and named according to the following: 1) the types and arrangement of cells in the tissue; 2) the type of ground substance composing the matrix; and 3) the type and arrangement of fibers in the matrix.

A | Cell Types

The following cell types may be found in some connective tissues.

1 | ***Fibroblasts:*** Responsible for producing the fibers and the ground substance of the connective tissue matrix.

2 | ***Macrophages:*** "Janitors" of the body; phagocytize foreign materials in the tissues.

3 | ***Mast cells:*** Produce histamine which is involved in allergic reactions and tissue inflammation.

4 | ***Adipocytes:*** Fat cells which store the body's triglycerides (fat).

5 | Other specialized cells may also be present, some of which are unique to a particular type of connective tissue. These will be discussed with the specific connective tissues.

Lab Tip: You will **never** find epithelial cells in connective tissue.

B | Types of Fibers in the Matrix

The following fibers may appear in abundance or not at all in specific connective tissues.

1 | ***Collagen fibers***

- These are the most common of the three fiber types.
- The individual fibers are long and straight (or slightly wavy), but they often occur in bundles.

- Collagen fibers are sometimes called ***white fibers*** because when they are present in large numbers in living tissues, the tissue looks white.
- Collagen fibers are composed of the protein collagen which is characterized by being ***very strong.***
- Bone, cartilage, tendons, and ligaments have large quantities of collagen fibers in them.

2 | ***Reticular fibers***

- Reticular fibers are much thinner than collagen fibers and tend to form branching networks with a lacey, dark-staining appearance.
- These fibers are also composed of the protein collagen.
- Reticular fibers provide a strong supporting net for many organs of the body (e.g., capillaries, nerves). They are also present in the basement membrane of epithelial tissue.

3 | ***Elastic fibers***

- Elastic fibers are thin and branching fibers which have the ability to stretch and return to original size and shape (this characteristic is known as elasticity).
- Elastic fibers are composed of the protein elastin.
- These fibers are found in the air sacs of the lungs, skin, and blood vessels.
- Elastic fibers are sometimes known as yellow fibers due to their color in living (fresh) tissue.

Some connective tissues are divided into two categories dependent upon the density (or closeness) of the protein fibers in the matrix.

- **Loose** connective tissues contain fibers which are not packed tightly together.
- **Dense** connective tissues contain fibers which are more tightly packed together.

C | *Type of Ground Substance as Part of the Matrix*

The ground substance is part of the non-living matrix. It is composed of a variety of chemicals. Embedded in the ground substance are the fibers and cells of the connective tissue.

Dependent upon the particular chemicals composing the ground substance, the matrix may exhibit the following characteristics.

1 | ***Hard, calcified:*** Found in bone. The hardness is due to the mineral salts found in the matrix.

2 | ***Rigid, somewhat flexible:*** Found in cartilage.

3 | ***Thick, viscous:*** Common in areolar connective tissue.

4 | ***Fluid or liquid:*** Such as the plasma of blood.

Lab Activity

Obtain slides for each of the following types of connective tissues. Using your textbook and other reference materials, be able to identify each of the different connective tissues. If you choose, you may make your own illustrations in the boxes provided.

Types of Connective Tissues

1 | Areolar connective tissue

- **Description:** This is the most widely distributed of all of the connective tissues. It has a variety of cells (mostly fibroblasts, macrophages) which are widely spaced and not arranged in a regular manner.

  There are many different fibers randomly scattered through the matrix which has a viscous, syrupy ground substance.

  Overall, the tissue looks very disorganized and messy. It is loosely arranged (not organized).

- **Function:** This tissue is located between epithelial tissue and muscle tissue and acts like a living "glue" holding these two other tissues together. It also forms thin membranes which help bind organs together.

- **Location:** Found between skin and its underlying structures (in the subcutaneous layer of skin), in mucous membranes, and around blood vessels, nerves, and body organs.

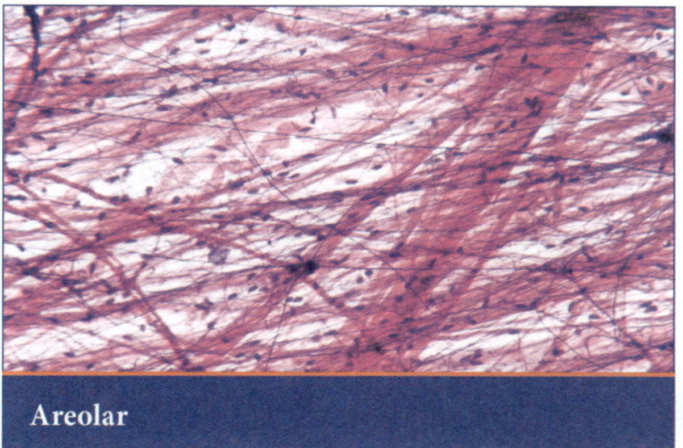

Areolar

Areolar drawing

2 | Adipose connective tissue

- **Description:** This is a type of connective tissue which contains mostly fat (adipose) cells.

  These fat cells (***adipocytes***) consist of a small area of cytoplasm and a large fat droplet.

  Adipose cells are connected very closely to each other. The cytoplasm and nucleus have been pushed to one side of the cell by the presence of the large fat droplet.

- **Function:** Adipose tissue helps insulate the body, protects (cushions) organs, stores energy in the form of fat, and shapes the body.

- **Location:** Adipose tissue is found under the skin and around certain organs of the body.

- **Special notes:** In some slides, the adipose tissue is clear (don't confuse it with simple squamous epithelial tissue), and in some slides the adipose droplet has been stained so that it can be seen. If available, examine ***both*** types.

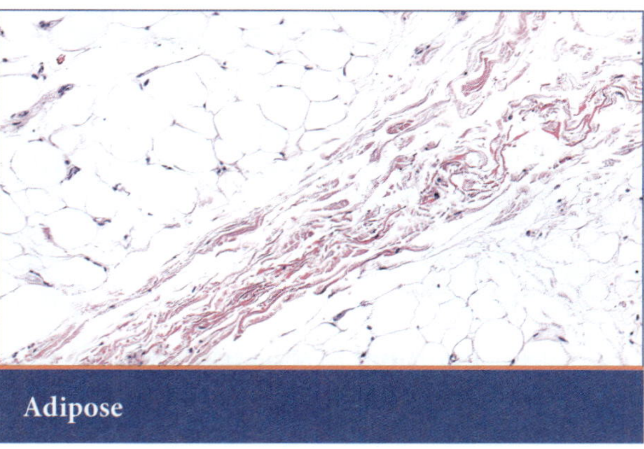

Adipose

Adipose drawing

3 | Reticular connective tissue

- **Description:** Composed primarily of reticular fibers which interconnect and branch yet are strong.

  In this slide you will see many small, stained cells with the dark reticular fibers scattered throughout. The reticular fibers are short and thin.

  This slide has so many cells that it looks like it has the measles or chickenpox.

  Some of these cells are reticular cells and others are a type of white blood cell known as a lymphocyte.

- **Function:** Defense and support.

- **Location:** This tissue type provides a framework to support the spleen, lymph nodes, and red bone marrow.

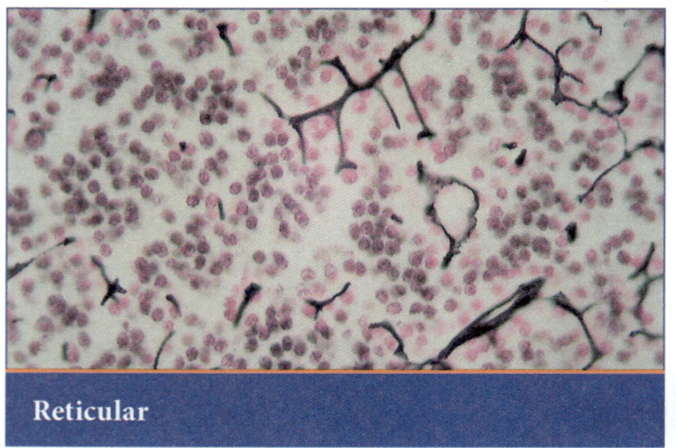

Reticular

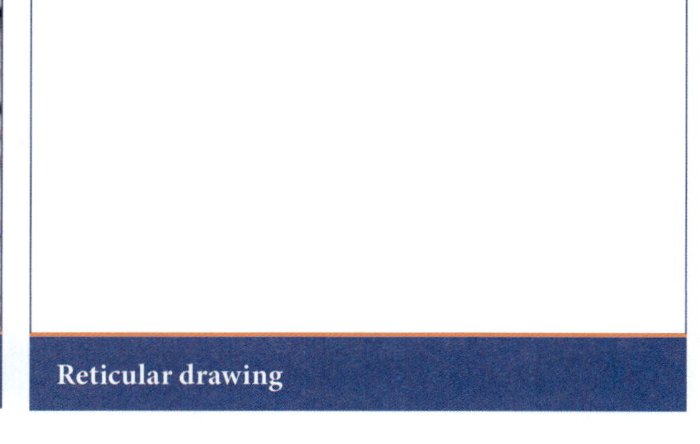

Reticular drawing

4 | Dense regular connective tissue

- **Description:** This tissue contains very few cells (a few fibroblasts which produce the many collagen fibers).

  This tissue is composed primarily of bundles of collagen fibers, which are usually arranged in orderly, parallel rows (looks wavy) with fibroblasts in between the rows. The fibers may appear "frayed."

  The term ***regular*** refers to the fact that the fibers in the tissue run in the same direction (parallel to the direction of pull). The term ***dense*** means that the numerous fibers are closely or densely packed together.

- **Function:** The arrangement of densely packed collagen fibers gives the tissue great strength and makes the tissue look silvery white in living tissue.

- **Location:** It is the tissue that composes tendons (attach muscle to bone), ligaments (attach bone to bone) and aponeuroses (sheetlike structures that attach muscle to muscle or muscle to bone).

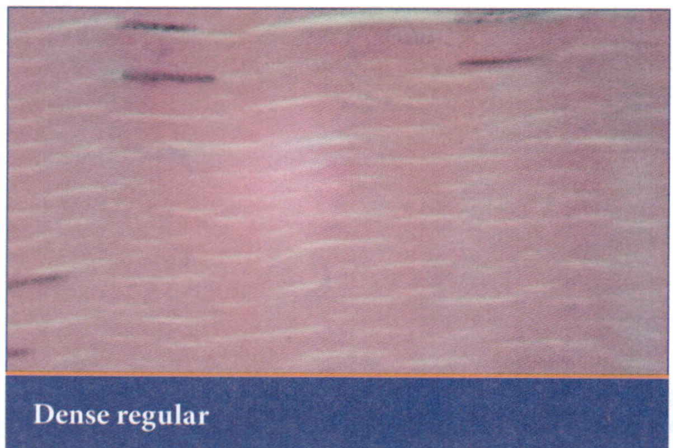

Dense regular

Dense regular drawing

5 | Dense irregular connective tissue

- **Description:** This tissue also contains collagen fibers but they are not arranged in an orderly fashion. The fibers are scattered in many different directions. The random arrangement of the collagen bundles accounts for the term ***irregular.***
- **Function:** The irregular arrangement of collagen fibers provides strength when a structure is being pulled or stressed in/from many directions.
- **Location:** This tissue is found in the deep layers of the skin, around bone and cartilage and some organs, and in some joints.

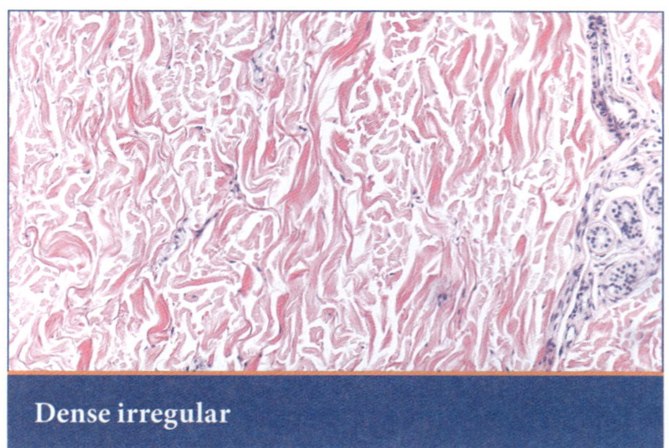

Dense irregular

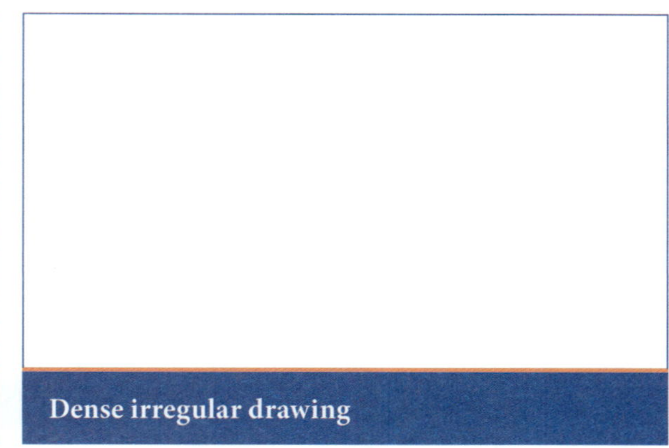

Dense irregular drawing

6 | Hyaline cartilage connective tissue

- **Description:** Hyaline cartilage is a hard, strong connective tissue, which contains collagen fibers in a firm yet flexible matrix. Even though collagen fibers are part of the structure of hyaline cartilage, they are often not apparent in slides. So, this tissue has a smooth appearing matrix.

  The cartilage contains many special cartilage cells known as ***chondrocytes*** which are located in cavities (spaces) called ***lacunae***.

  In many slides, the chondrocytes appear to be stained pink to violet. The lacunae usually appear clear.

  In living tissue, hyaline cartilage would appear as a shiny, bluish-white tissue. It is sometimes known as gristle.

  Cartilage is a connective tissue which doesn't contain blood vessels or nerve fibers.

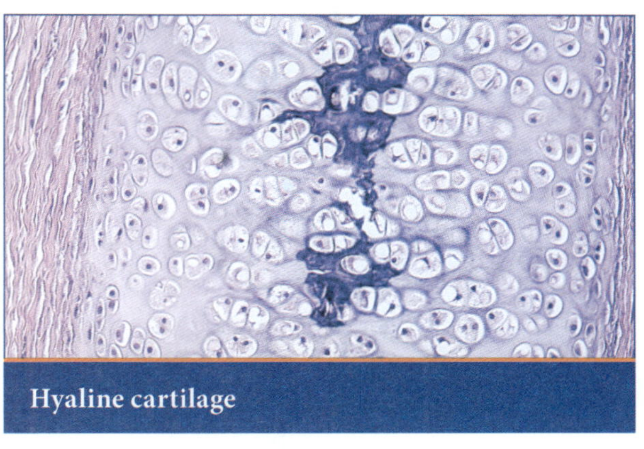

Hyaline cartilage

Hyaline cartilage drawing

- **Function:** Hyaline cartilage helps provide smooth movement at some joints; it also is involved in connecting some bones and in support.
- **Location:** This is the most common type of cartilage in the body and can be found at many joints, at the ends of ribs, in parts of the respiratory system, and as part of the structure of the outer nose.

7 | Bone connective tissue

- **Description:** Bone connective tissue consists of special bone cells (known as ***osteocytes***) which are also found in lacunae.

  The matrix of bone is hard and calcified. It does contain collagen fibers which provide strength.

  There are two types of bone—compact and spongy. The type you will examine in lab is compact bone.

  Compact bone has a very characteristic structure. It consists of many ***osteons*** (or ***Haversian systems***) packed tightly together.

  In the center of each osteon is a central canal, which usually appears clear in the microscope slides. The rest of the osteon resembles the rings of a tree trunk. These concentric rings are known as ***lamellae***.

- **Function:** Support, protection, fat storage, and production of blood.

- **Location:** Bones of the body.

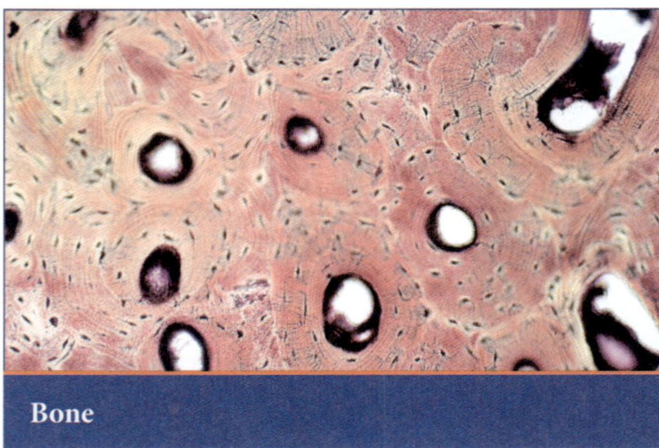

Bone

Bone drawing

8 | Blood connective tissue

- **Description:** This tissue consists of a liquid matrix (blood plasma) in which there are the blood cells (red blood cells, white blood cells, and platelets).

  Blood slides will show numerous, very small, pink-stained red blood cells. Interspersed throughout the slide will be a few white blood cells which are usually stained purple or blue and which have distinctly darker-stained nuclei. You may also notice platelets in the slide; platelets are significantly smaller than the other cells and are darkly stained.

- **Function:** The function of blood is to transport materials from one part of the body to another and to aid in the body's defense mechanisms. It also is involved in maintaining body temperature and chemistry.

- **Location:** Blood vessels and heart chambers.

- **Special notes:** It can be difficult to focus the blood slide because of the small size of the blood cells. Please ask your instructor for assistance if you have difficulty focusing this slide.

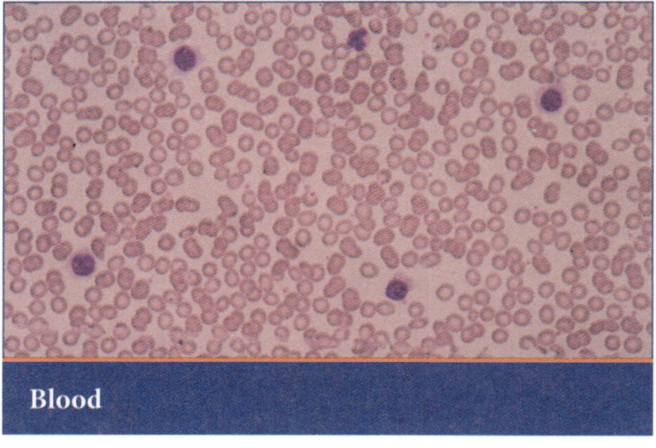

Blood

Blood drawing

Muscle Tissue (Three Types to Identify)

Muscle tissue is highly specialized for contraction to produce movement of the body, body parts, or materials through organs. Muscle tissue also helps maintain posture and is involved in producing heat (as it contracts).

There are three different types of muscle tissue in the human body.

Lab Activity

Obtain slides for each of the following types of muscle tissues. Using your textbook and other reference materials, be able to identify each of the different muscle tissues. If you choose, you may make your own illustrations in the boxes provided.

Types of Muscle Tissue

1 | **Skeletal muscle tissue**

- **Description:** Skeletal muscle consists of many muscle fibers (cells) which are long, cylindrical, slender, and are arranged in parallel bundles.

  Skeletal muscle fibers may contain many nuclei per cell, and the nuclei tend to be located near the edges of the fibers.

  The fibers of skeletal muscle tissue are ***striated*** (striped dark and light) in appearance. This striping may be ***very*** light so proper focusing and adjustment of the light source is absolutely necessary.

  You may see adipose tissue interspersed between the bundles of these muscle fibers.

- **Function:** Contraction which produces movement of the body or its parts, posture, and heat production. It is voluntary tissue (under conscious control).

- **Location:** Attached to bones.

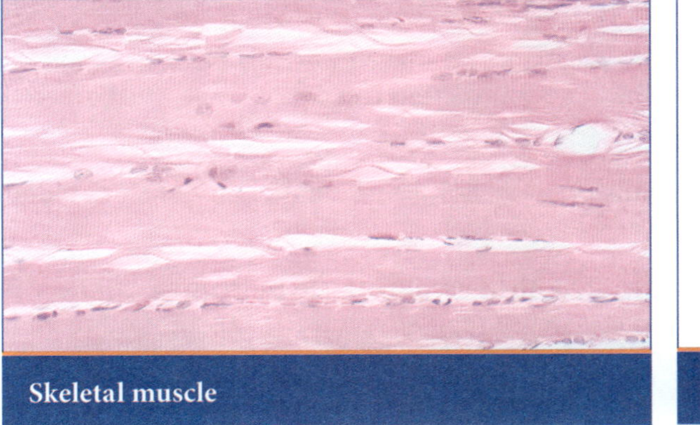

Skeletal muscle

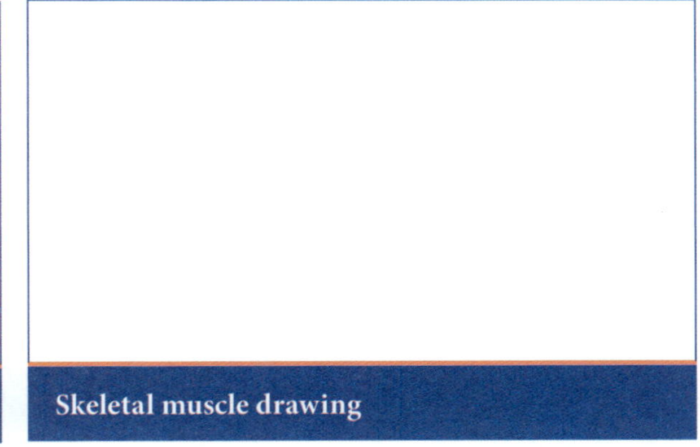

Skeletal muscle drawing

2 | **Cardiac muscle tissue**

- **Description:** Cardiac muscle cells are shorter than skeletal muscle fibers, and they branch and interconnect.

  Similar to skeletal muscle fibers, the cells of cardiac muscle are also striated. However, the striations in these cells tend to be even less distinct than in skeletal muscle. Look closely to observe them.

  Cardiac muscle cells may contain 1–2 centrally-located nuclei per cell, and the nuclei are usually more evident than in skeletal muscle tissue.

Where the adjacent cells join together, there will be a dark band known as the ***intercalated disc***. The presence of these dark bands will allow you to more easily distinguish this tissue from skeletal muscle tissue. These discs allow cardiac muscle cells to communicate with each other so that the cells contract as a unit when the heart beats.

- **Function:** Movement of blood through the body. Cardiac muscle is an involuntary tissue (not under conscious control).
- **Location:** This tissue is found in only one place—the walls of the heart.

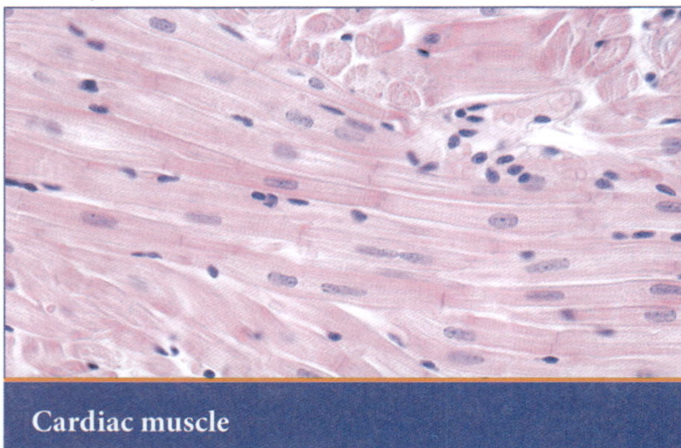
Cardiac muscle

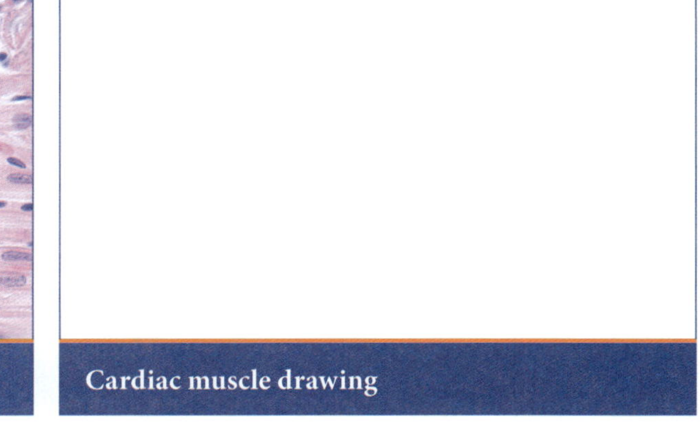
Cardiac muscle drawing

3 | **Smooth muscle tissue**

- **Description:** The cells of smooth muscle are ***not*** striated.

  The cells tend to be football-shaped (fat in the middle and tapering towards the ends) although in many slides this shape is difficult to observe.

  Smooth muscle cells contain one nucleus per cell. It is usually an elongated nucleus located in the center of the cell.

  Smooth muscle cells are generally arranged in sheets of muscle tissue.

  Smooth muscle is involuntary.

- **Function:** The role of smooth muscle is to contract and move materials (such as blood, food, and urine) through the organs where it is found. The contraction can also change the diameter of some of the organs.
- **Location:** This tissue is located in the walls of many hollow internal organs such as blood vessels, digestive system organs, uterus, and the urinary bladder.
- **Special notes:** Initially, you may confuse smooth muscle with dense regular connective tissue. However, the smooth muscle is not wavy as is dense regular connective tissue, and there is a more regular distribution of nuclei in smooth muscle tissue. In addition, you may look for the "frayed" fibers that would indicate dense regular connective tissue.

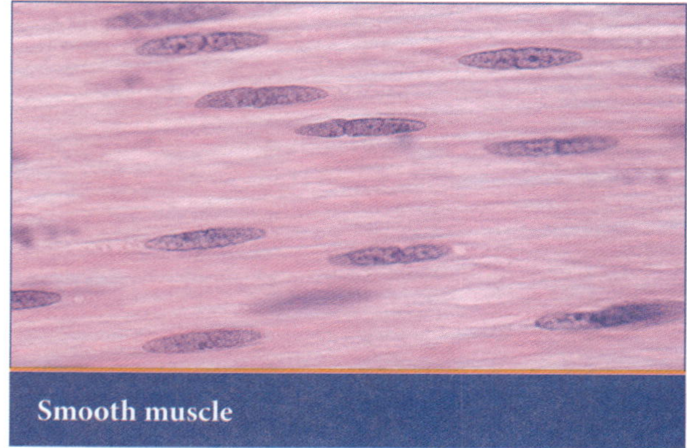
Smooth muscle

Smooth muscle drawing

Nervous Tissue (One Type to Identify)

Nervous tissue is another highly specialized tissue. It has one role in the body—to send and receive nerve impulses. Nerve impulses are used for communication and controlling body functions.

Nervous tissue is found in the brain, spinal cord, and nerves.

General Characteristics of Nervous Tissue

- Nervous tissue consists of two cell types—***neurons*** and ***neuroglia.*** Neurons are the cells in nervous tissue which are specialized to send and receive nerve impulses. They are composed of several components—a large central ***cell body*** with a well-defined nucleus and many slender, thread-like projections which are called ***axons*** and ***dendrites.***

  Neuroglia are cells in the nervous tissue which are smaller and more numerous than neurons. They function to protect and support the neurons but they ***do not*** transmit or receive nerve impulses.

Lab Activity

Obtain slides for nervous tissue. Using your textbook and other reference materials, be able to identify nervous tissue. If you choose, you may make your own illustrations in the box provided.

1 | **Nervous tissue**

- **Description:** The slide of nervous tissue (giant multipolar neuron) should show several large neurons with their darkly stained nuclei. You should be able to see the slender projections extending from the cell body.

  The numerous neuroglia will appear as small dark dots on the slide.

- **Function:** React to stimuli, send and receive nerve impulses.

- **Location:** Nervous system.

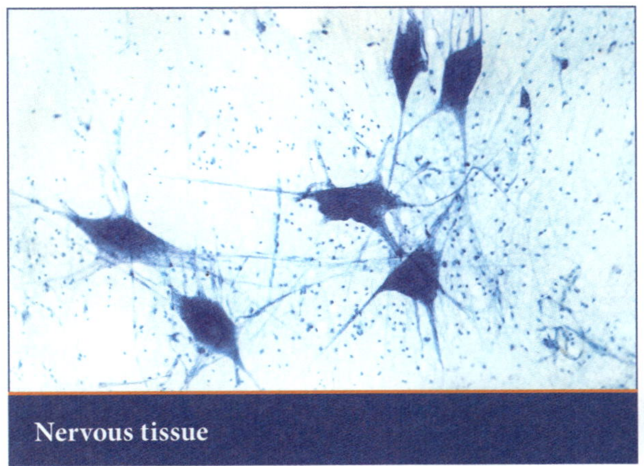

Nervous tissue

Nervous tissue drawing

Summary of Tissues for Lab Practical

Epithelial Tissues

1 | Simple squamous epithelial tissue

2 | Stratified squamous epithelial tissue

3 | Simple cuboidal epithelial tissue

4 | Simple columnar epithelial tissue, ciliated or nonciliated

5 | Ciliated pseudostratified columnar epithelial tissue

Connective Tissues

1 | Areolar connective tissue

2 | Adipose connective tissue

3 | Reticular connective tissue

4 | Dense regular connective tissue

5 | Dense irregular connective tissue

6 | Hyaline cartilage connective tissue

7 | Bone connective tissue

8 | Blood connective tissue

Muscle Tissues

1 | Skeletal muscle tissue

2 | Cardiac muscle tissue

3 | Smooth muscle tissue

Nervous Tissue

1 | Nervous tissue

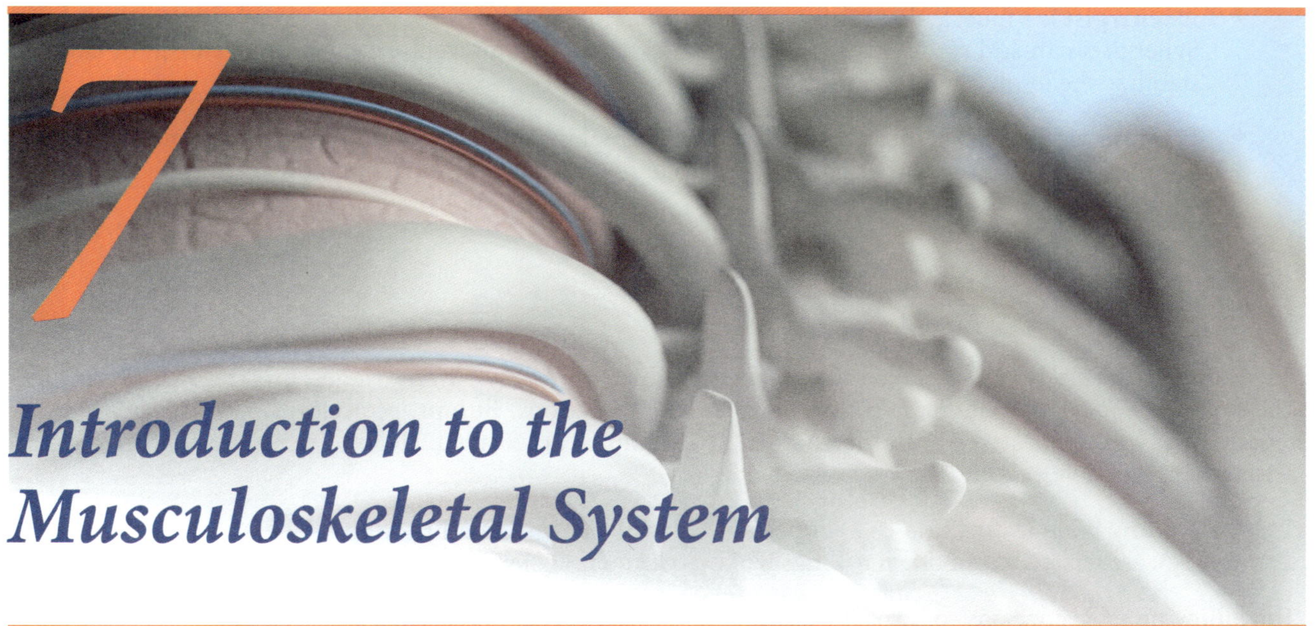

7 Introduction to the Musculoskeletal System

How the Body Produces Movement

In the laboratory, you will be studying the bones and skeletal muscles as a unit and will examine how movement is produced by the interaction between the bones and muscles.

Generally, muscles have attachment points to at least two bones, and the muscle must generally cross a joint in order for movement to occur. Skeletal muscle attaches to bone via tendons. When a muscle contracts, it typically shortens. This places force on the insertion tendon and allows the bone to move.

The **origin** is where the muscle attaches to a non-moving or stationary bone when a muscle contracts. The origin acts as an anchor. The **insertion** is the attachment of the muscle to the bone that moves when the muscle contracts. As a muscle shortens, the insertion will move towards the origin. The bulky (meaty) portion of the muscle in between the origin and insertion is the **belly.**

For complex, coordinated movements, muscles generally don't act alone but work together as a group. In many cases, muscles are found in **opposing (antagonistic) pairs** which means that whatever movement one muscle produces when it contracts, there will be a muscle on the opposite side that will produce an opposite movement when it contracts. The elbow joint is a good example of this concept. When the biceps brachii muscle (on the anterior part of the upper arm) contracts, it flexes the elbow; but when the triceps brachii muscle (on the posterior part of the upper arm) contracts, it results in extension (straightening) of the elbow joint. Thus, the biceps brachii and triceps brachii are opposing (antagonistic) muscles.

When groups of muscles work together to produce movement, each muscle in that group is classified in one of four categories, based on its role in the movement:

- The **prime mover** or **agonist** is the muscle responsible for producing the desired movement when it contracts. If you wanted to flex your elbow, your biceps brachii muscle would be an agonist.
- An **antagonist** is a muscle that produces an action opposite the agonist. It is usually stretched or partly relaxed in order for the agonist to function properly. To flex your elbow, your triceps brachii muscle acts as the antagonist.

  The roles of agonists and antagonists can change. Using our elbow example, if you wanted to extend (straighten) your elbow, the triceps brachii muscle would contract and would be known as the agonist. But, the biceps brachii muscle would need to stretch or relax in order to straighten the elbow, so it is now known as the antagonist!

- A **synergist** is a muscle that contracts at the same time as the agonist and acts as a helper to the agonist. Synergists can add some extra force to the movement, can prevent unwanted movement, or can help stabilize the joint.
- **Fixators** are synergists that act specifically to immobilize ("fix") a bone at or near the muscle's origin. They help to steady the proximal end of a bone so that movement occurs more effectively at the distal end.

Learning Bone Parts

As you begin the study of bones, you will notice that there are many bumps, ridges, protrusions, depressions, and holes in bones. These bone markings have specific functions. Bumps, projections, and ridges are where tendons and ligaments attach and where joints (articulations) are formed. Holes, depressions, and grooves are areas where blood vessels and nerves are located.

There are ***many*** bone features on each bone; you will only learn a small portion. Understanding some of the terminology associated with bone markings will help immensely in the study of bones.

The following table contains some of the terminology you will encounter while studying bones.

Table 7.1 | Bone Marking Terminology

General Description	Term	Definition/Description
Elevations and projections that form attachment points for tendons or ligaments	Crest	Prominent ridge or elongated projection
	Epicondyle	Rough projection (bump) above a condyle
	Linea/line	Long, narrow, low ridge or border; sometimes rather faint
	Process	Prominent raised area or projection
	Spine or spinous process	Sharp or pointed, slender projection or narrow process
	Trochanter	Large, rough projection (only found on a femur)
	Tubercle	Small, rounded bump (smaller than a tuberosity)
	Tuberosity	Large, rounded, usually rough bump
Processes/markings that form joints (articulations)	Condyle ("knuckle")	Large, smooth, rounded process at the end of a bone; usually fits into a fossa to form a joint
	Facet	Small smooth, flat articular surface
	Head	Rounded, expanded end of an epiphysis; separated from the shaft by a narrow neck
	Neck	Constricted, narrow connection between epiphysis and diaphysis; usually at the base of a head
	Trochlea ("pulley")	Smooth, grooved articular process
Depressions	Fossa ("trench")	Shallow depression (valley); often part of a joint
	Sulcus/Groove	Narrow depression for a tendon, blood vessel, or nerve
Openings	Canal/Meatus	Tube-like opening; tunnel; passageway
	Fissure	Elongated, narrow crack-like slit for passage of blood vessels and nerves
	Foramen	Rounded hole for passage of blood vessels and nerves
	Sinus	Chamber or cavity within a bone, normally filled with air

Naming Muscles

Learning muscle names can seem just as confusing as bone marking terminology. There are several characteristics that are used in naming muscles. Learning some of these terms will guide you in remembering the muscles.

Table 7.2 | Characteristics Used to Name Muscles

General Description	Term	Meaning	Example
Location	Femoris	Femur	Biceps femoris
	Oris	Mouth	Orbicularis oris
	Temporalis	Temples	Temporalis
	Tibialis	Tibia	Tibialis anterior
Relative Size	Brevis	Shortest	Adductor brevis
	Longus	Longest	Adductor longus
	Maximus	Largest	Gluteus maximus
	Minimus	Smallest	Gluteus minimus
	Major	Larger	Pectoralis major
	Minor	Smaller	Pectoralis minor
	Latissimus	Widest	Latissimus dorsi
	Magnus	Large	Adductor magnus
	Vastus	Huge	Vastus medialis
Shape	Deltoid	Triangular	Deltoid
	Trapezius	Trapezoid	Trapezius
	Serratus	Saw-toothed	Serratus anterior
	Orbicularis	Circular	Orbicularis oculi
Direction of Fibers	Rectus	Parallel to midline; straight	Rectus abdominis
	Transverse (transversus)	Perpendicular to midline	Transverse abdominis
	Oblique	Diagonal to midline	External oblique
Number of Origins	Biceps	Two origins	Biceps brachii
	Triceps	Three origins	Triceps brachii
	Quadriceps	Four origins	Quadriceps femoris
Origin and Insertion	Sternocleidomastoid	Origin on sternum and clavicle; insertion on mastoid process of temporal bone	Sternocleidomastoid
	Stylohyoid	Origin on styloid process of temporal bone; insertion on hyoid bone	Stylohyoid
Function/ Action	Flexor	Decreases an angle at a joint; bends	Flexor carpi ulnaris
	Extensor	Increases an angle at a joint; straightens	Extensor carpi radialis longus
	Abductor	Movement away from the midline	Abductor pollicis longus
	Adductor	Movement towards the midline	Adductor longus
	Levator	Raises (elevates) a body part	Levator scapulae

8 The Musculoskeletal System—Lower Extremity

Lab Activity 1 | Bones of the Pelvic Girdle and Leg

From the bone cases provided, take out the bones of the pelvic girdle, leg, and foot only. Using your textbook and other references as guides, identify the following bones and bone features.

A | Pelvic Girdle

1 | **Coxa** (os coxa)

- Ilium
- Iliac crest
- Iliac fossa
- Anterior superior iliac spine
- Anterior inferior iliac spine
- Posterior superior iliac spine
- Posterior inferior iliac spine
- Greater sciatic notch
- Ischium
- Ischial tuberosity
- Pubis
- Acetabulum (part of hip joint)
- Obturator foramen (for passage of obturator nerve & vessels to the thigh)
- Pubic symphysis (where two pubic bones meet)

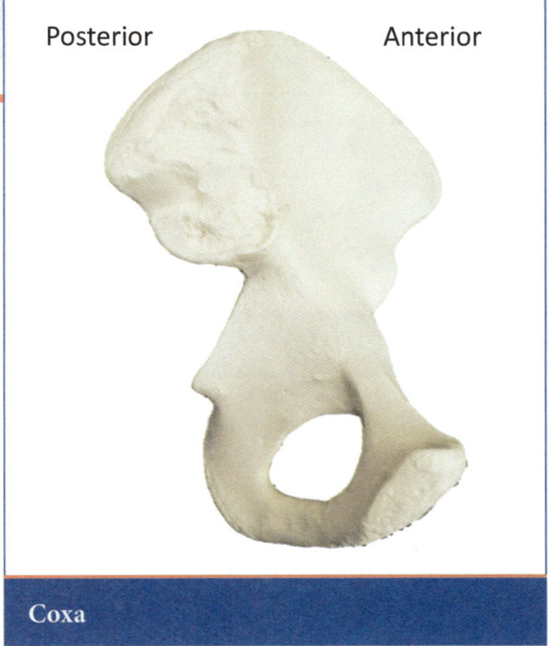

Coxa

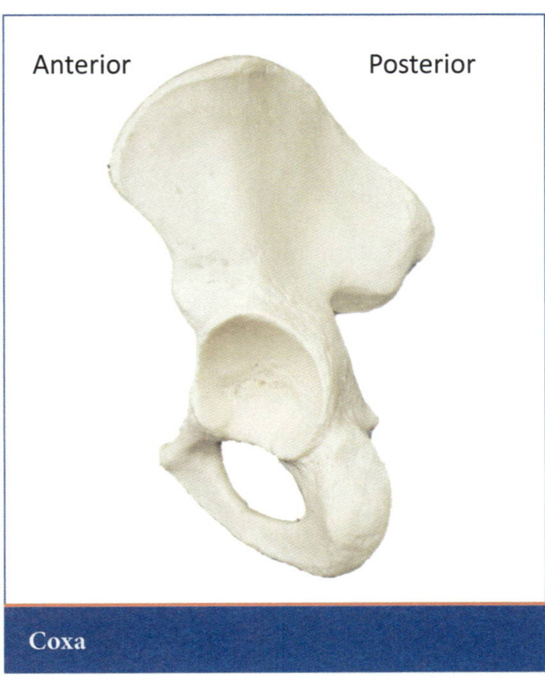

Coxa

B | Lower Extremity

1 | **Femur**

- Head (part of hip joint)
- Neck
- Greater trochanter
- Lesser trochanter
- Gluteal tuberosity
- Linea aspera
- Medial condyle
- Lateral condyle

2 | **Patella**

Patella

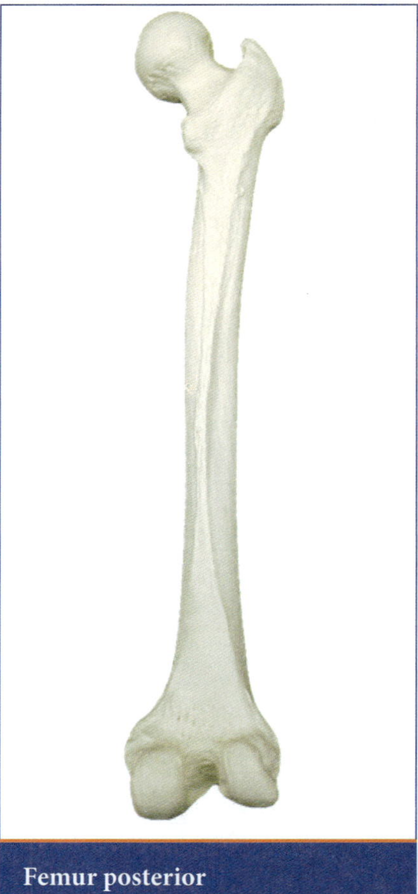

Femur posterior

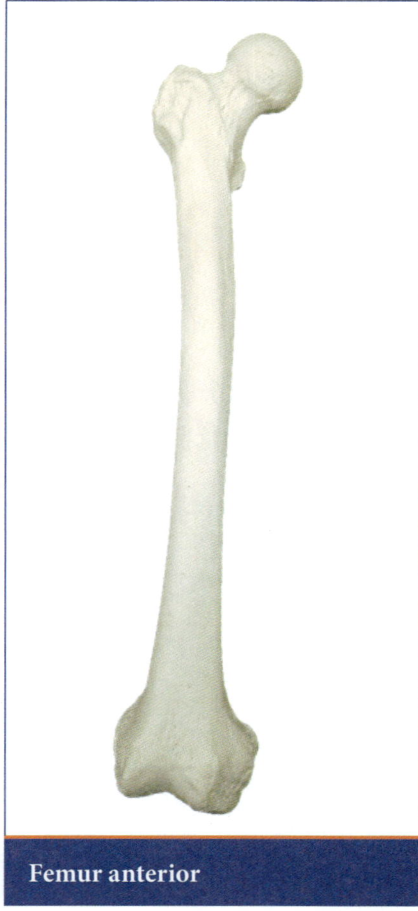

Femur anterior

3 | **Tibia**

- Anterior border (margin) (crest)
- Lateral condyle
- Medial condyle
- Tibial tuberosity
- Medial malleolus

4 | **Fibula**

- Head
- Lateral malleolus

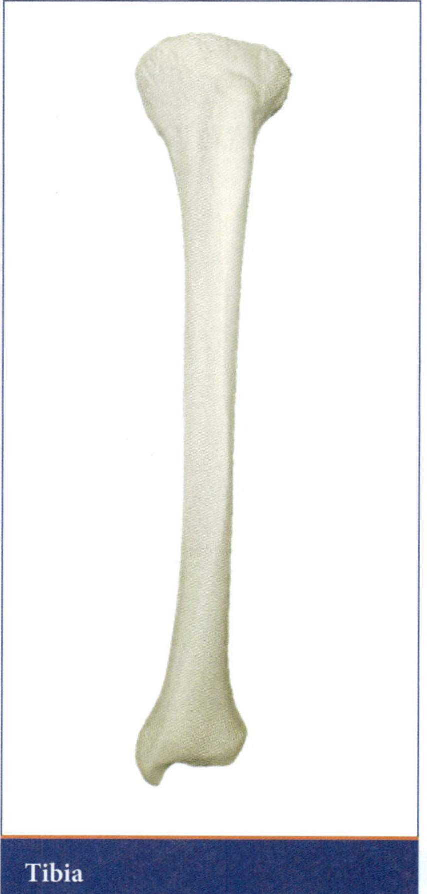

Tibia

Fibula

5 | **Tarsals**

- Calcaneus
- Talus
- Cuboid
- Navicular
- Cuneiforms—1st, 2nd, 3rd (also known as medial, intermediate, and lateral)

(**C**ome **T**o **C**uba **N**ext **C**hristmas)

6 | **Metatarsals 1–5 (must know number)**

7 | **Phalanges 1–5 (must know number)**

- Proximal phalanx
- Middle phalanx
- Distal phalanx

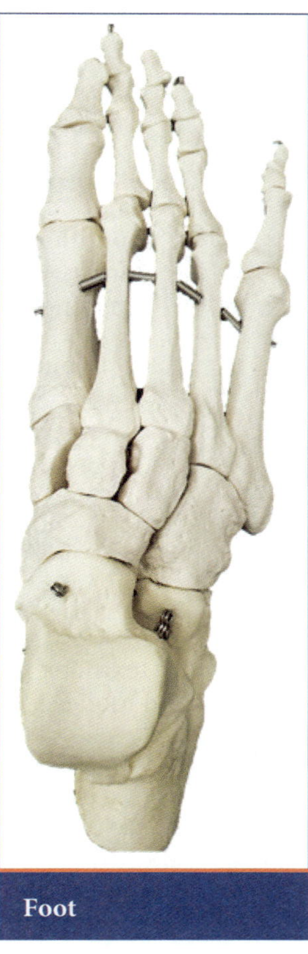

Foot

Lab Activity 2 | Muscles of the Pelvic Girdle and Leg

Muscles That Move the Femur (Thigh) at the Hip Joint and Move the Leg at the Knee Joint

The hip joint (a ball-and-socket joint) permits a wide range of movements including flexion, extension, abduction, adduction, rotation, and circumduction. Most muscles that move the thigh originate on the pelvic girdle and insert on the femur. The muscles that move the thigh tend to be some of the most powerful muscles in the body.

- **Flexors and extensors:** In order to affect movement in both the hip and knee joint, the muscles must cross both joints. Usually the anterior muscles of the thigh tend to flex the femur at the hip and extend the tibia at the knee (this is the motion your leg takes when it is making the forward movement in walking); the posterior muscles tend to extend the femur at the hip joint and flex the tibia at the knee (the type of motion your leg takes when it is making the backswing motion in walking).
    - **Flexors:** The hip (thigh) flexors lie anterior to the hip joint.
    - **Extensors:** The hip (thigh) extensors lie posterior to the hip joint. The massive hamstring group is a primary extensor of the femur at the hip. During forceful extension, the gluteus maximus muscle is also activated. The hamstrings cross both the hip and knee joints so they not act on both the thigh and leg. The term "hamstring" originated because of the practice of butchers who used the tendons of these muscles to hang their hams for smoking. "Pulled hamstrings" are common sports injuries in sprinting athletes.
- **Abductors and adductors:** The abductors and adductors of the thigh are very important muscles in walking. They help to keep the body's weight balanced over the limb which is on the ground.
    - **Abductors:** The abductors tend to be located on the lateral surface of the thigh and originate on the ilium.
    - **Adductors:** The adductors consist of a variety of muscles located on medial surfaces of the thigh. The adductor muscles originate on the inferior part of the pubis. Adductors are used in movements that press the thighs together as when riding a horse. They are important in walking movements and in fixing the hip when the knee is flexed. A "pulled groin" refers to straining or stretching of these muscles.

Muscles That Move the Foot

There are a wide variety of movements allowed in the joints of the foot, including dorsiflexion and plantar flexion, inversion and eversion, and flexion and extension.

- **Dorsiflexors and plantar flexors**
  - **Dorsiflexors:** The muscles in the anterior part of the leg are primarily toe extensors and ankle dorsiflexors. Dorsiflexion is not a very powerful movement, but it is needed to keep the toes from dragging during walking.
    - **Plantar flexors:** Muscles located laterally are involved in plantar flexion and eversion. The posterior muscles work primarily to plantar flex the foot and flex the toes. Plantar flexion is a very powerful movement involving the ankle and foot joints because it lifts the entire weight of our bodies. It provides the forward drive in walking and running and allows us to stand on our tip-toes.

Lab Activity

Using your textbook and other reference materials as guides, locate the following muscles on the human cadavers and models. You will be responsible for the origins, insertions, and actions as well as identification.

Table 8.1 | Anterior Muscles

Muscle Name	Origin (O) and Insertion (I)	Action
You need to be able to **identify** and **correctly spell** each muscle.	For each muscle, you **are required to** learn ***all*** of the origins and insertions listed below. Any items in parentheses ( ) may be optional based upon your individual instructor's preferences.	You **are required to** learn the action(s) indicated in **bold** type. The remainder of the material concerning actions is for informational purposes only.
Muscles that Flex the Hip and Move the Knee		
Iliopsoas: (This is actually two muscles, the iliacus which is lateral and the psoas major which is medial. The psoas is called the tenderloin by butchers.)	**O:** Iliac fossa (iliacus); lumbar vertebrae (psoas) **I:** Lesser trochanter	**Flexes hip.**
Quadriceps femoris group: Consists of four muscles located on the front and sides of the thigh.	All of these muscles have the same insertion: **I:** Tibial tuberosity (via patella)	
Rectus femoris	**O:** Anterior inferior iliac spine	**Flexes the hip and extends the knee.**
Vastus lateralis	**O:** Greater trochanter	**Extends knee.**
Vastus intermedius	**O:** Anterior surface of proximal femur	**Extends knee.**
Vastus medialis	**O:** Linea aspera	**Extends knee.** These muscles are powerful knee extensors and are used for climbing, jumping, running, and rising from the sitting position. The quadriceps group plays an important role in strengthening the knee joint. The inferior fibers of the vastus medialis muscle also help to stabilize the patella.

Sartorius	**O:** Anterior superior iliac spine **I:** Medial side of tibia	**Flexes hip and knee; laterally rotates the hip.** These are the motions used in crossing the leg.
	Muscles that Abduct and Adduct the Hip	
Tensor fasciae latae	**O:** Iliac crest **I:** Lateral side of tibia (via the iliotibial tract)	**Abducts hip.** Synergist of gluteus medius muscle. It pulls the iliotibial tract tight which plays an important role in standing by helping to keep the knee extended.
Gracilis	**O:** Pubis **I:** Medial side of tibia	**Adducts hip.**
Adductor longus	**O:** Pubis **I:** Linea aspera	**Adducts hip.**
Pectineus	**O:** Pubis **I:** Proximal linea aspera	**Adducts hip.**
	Muscles that Move the Foot	
Tibialis anterior	**O:** Anterior and lateral surfaces of proximal tibia **I:** 1st/medial cuneiform; 1st metatarsal	**Dorsiflexion and inversion.**

Table 8.2 | Posterior Muscles

Muscle Name	Origin (O) and Insertion (I)	Action
You need to be able to **identify** and **correctly spell** each muscle.	For each muscle, you **are required to** learn ***all*** of the origins and insertions listed below. Any items in parentheses ( ) may be optional based upon your individual instructor's preferences.	You **are required to** learn the action(s) indicated in **bold** type. The remainder of the material concerning actions is for informational purposes only.
	Muscles that Extend the Hip and Flex the Knee	
Hamstring group—Consists of three muscles: **Biceps femoris** (most lateral) **Semitendinosus** (intermediate) **Semimembranosus** (most medial)	All three hamstring muscles have the same origin: **O:** Ischial tuberosity **I:** Head of fibula **I:** Medial side of tibia (near insertion of gracilis) **I:** Medial condyle of tibia	All hamstring muscles **extend** the **hip** and **flex** the **knee.** The actions of the hamstrings are opposite those of the quadriceps muscles.
Gluteus maximus	**O:** Dorsal (posterior) ilium and sacrum **I:** Gluteal tuberosity	**Extends hip** (primary role); also used in climbing stairs, running. Generally inactive during walking. Antagonist of iliopsoas muscle.

Muscles that Abduct the Hip		
Gluteus medius	**O:** Lateral surface of ilium **I:** Greater trochanter	**Abducts hip;** important in walking or running. This muscle steadies the pelvis so it doesn't tilt downward (sag) when the foot of the opposite side is taken off the ground and thrust forward. This action helps allow the forward-swinging foot to clear the ground.
Muscles that Laterally Rotate the Hip		
Piriformis	**O:** Anterior sacrum **I:** Greater trochanter	**Laterally rotates hip**
Gastrocnemius (Superficial posterior muscle of the lower leg that has two "bellies" that are well developed in ballet dancers and body builders.)	**O:** Medial and lateral condyles of femur **I:** Calcaneus (via calcaneal tendon)	**Plantar flexion.** The gastrocnemius and soleus muscles are responsible for helping to lift the heel off the ground and to provide forward propulsion during walking and running.
Soleus (Posterior muscle deep to the gastrocnemius.)	**O:** Proximal tibia & fibula **I:** Calcaneus (via calcaneal tendon)	**Plantar flexion.** Important during walking, running, and dancing.
Fibularis longus	**O:** Head of fibula **I:** 1st/medial cuneiform and 1st metatarsal. (The fibularis longus muscle extends laterally along the lower leg, then attaches to the insertions by a long tendon that curves under the foot.)	**Plantar flexion; eversion.** This muscle helps keep the foot flat on the ground.

9

Goniometry and Range of Motion—Lower Extremity

Introduction

The purpose of this lab is to determine the active range of motion allowed at several different joints. A joint, or an articulation, is where two or more bones meet. Gross movements of the body occur at these joints and are caused by contractions of skeletal muscles. Assessing joint range of motion (ROM) is an important step in understanding the overall health of the joint.

Goniometers (gonio = angle, meter = device to measure) are used to assess ROM. When used properly, the measurements taken using a goniometer are both reliable and valid. Goniometry is used mainly in health and fitness facilities especially physical therapy, orthopedic, and chiropractic offices to help assess injured patients. Goniometers are assessment tools and can be used to obtain a baseline measurement and to determine the effectiveness of treatment. Each joint is unique in its structure and in the amount of movement allowed. The anatomical adage of form equals function is never so obvious as it is in the articulations of the skeleton.

The goniometer has three parts:

- The **axis** is the point where the two arms of the goniometer meet. The arms will be able to rotate around this axis. We will be using this axis to ensure proper positioning of the goniometer.
- The **stationary arm** of the goniometer has a compass with numbers arranged in a circle. These numbers are read as degrees. As the name implies, this arm will remain fixed when taking a measurement.
- The **movable arm** will have a line and/or arrow at its axial end that will rotate around the compass of the stationary arm. This arm will move with the patient's limb and the ROM can be assessed by reading the numbers in **black** (not red) through which the line passes.

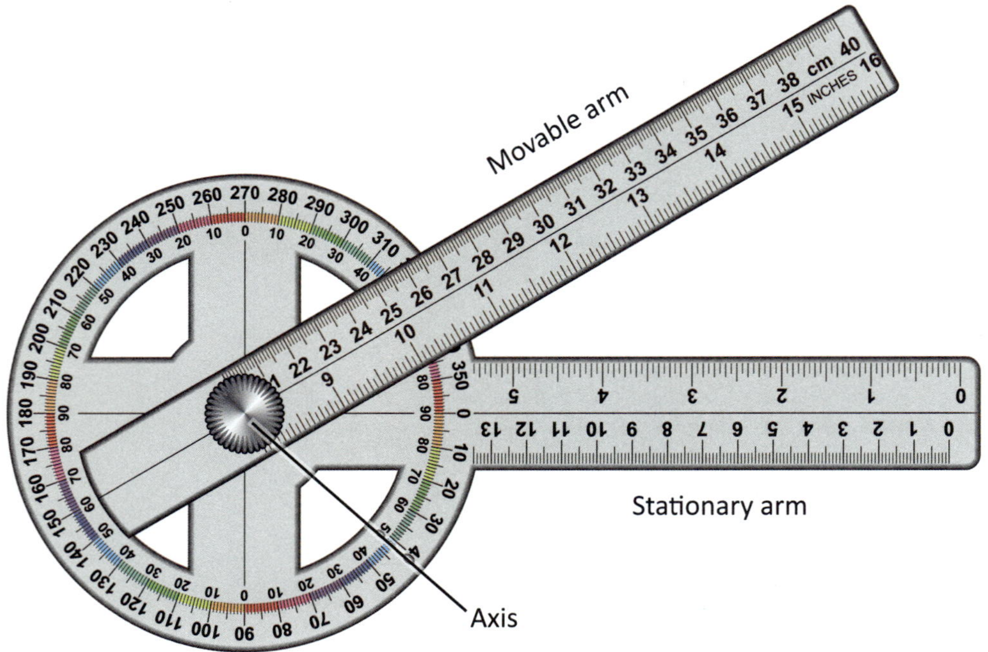

Figure 9.1 | Goniometer

Procedures

1 | Working with a lab partner, you will measure the ROM of each other's lower extremity joints.

2 | Each measurement will be taken two times, and the average calculated.

3 | When recording the measurement, use the **black numbers** on the goniometer.

4 | All data will be recorded in the tables.

Lab Note: In some clinical settings, goniometry measurements may be taken with the patient in a supine position.

Lab Activity 1 | Hip Joint

The proximal bone in this joint is the os coxa, specifically the acetabulum, and the distal bone is the femur. This is a synovial, ball and socket joint that will allow movement in all three planes of motion.

1 | **Flexion: 100° (normal measurement)**

a | Patient position

i | Standing with back against the wall.

1 | This will help stabilize the pelvis.

b | Goniometer position

i | **Axis:** Over the greater trochanter.

ii | **Stationary arm:** Oriented along the lateral side of the trunk (mid-axillary line).

iii | **Movable arm:** Oriented along the lateral side of the femur, pointing distally.

iv | The line of the movable arm should be aligned with the 0° on the compass of the stationary arm.

c | Directions

i | Without assistance, ask the patient to bring his/her knee up toward their chest. ***Try not to move the stationary arm!***

ii | Move the movable arm to realign with the femur.

iii | Look for the line of the movable arm as it passes through the compass. Record the measurement in Table 9.1.

Table 9.1 | Flexion: 100° (normal measurement)

	1st Measurement	2nd Measurement	Average
Student 1			
Student 2			

2 | **Extension: 30° (normal measurement)**

a | Patient position

i | Standing facing the wall.

b | Goniometer position

i | **Axis:** Over the greater trochanter.

ii | **Stationary arm:** Oriented up along the lateral side of the trunk (mid-axillary line).

iii | **Movable arm:** Oriented along the lateral side of the femur, pointing distally.

iv | The line of the movable arm should be aligned with the 0° on the compass of the stationary arm.

c | Directions

i | Without assistance, ask the patient to extend his/her thigh as much as possible. ***Try not to move the stationary arm!***

ii | Move the movable arm to realign with the femur.

iii | Look for the line of the movable arm as it passes through the compass. Record the measurement in Table 9.2.

Table 9.2 | Extension: 30° (normal measurement)

	1st Measurement	2nd Measurement	Average
Student 1			
Student 2			

3 | **Abduction: 40° (normal measurement)**

a | Patient position

i | Standing with back against the wall, or with the back supported against a lab table.

ii | Feet hip width apart.

b | Goniometer position

i | **Axis:** Over the anterior superior iliac spine (ASIS) of the limb being tested.

ii | **Stationary arm:** Pointing toward the opposite ASIS.

iii | **Movable arm:** Oriented along the anterior side of the femur.

iv | The line of the movable arm should be aligned with the 90° mark of the compass.

c | Directions

i | Without assistance, ask the patient to abduct the thigh as far as possible. ***Do not move the stationary arm.***

ii | Make sure that the patient's torso remains straight; the patient should only move the thigh.

iii | Move the movable arm to realign with the femur.

iv | Look for the line of the movable arm as it passes through the compass. Record the measurement in Table 9.3.

Table 9.3 | Abduction: 40° (normal measurement)

	1st Measurement	2nd Measurement	Average
Student 1			
Student 2			

4 | **Adduction: 20° (normal measurement)**

a | Patient position

i | Standing with back against the wall or supported next to a table.

ii | Place the feet so that they are the width of the hips.

b | Goniometer position

i | **Axis:** Over the ASIS of the limb being tested.

ii | **Stationary arm:** Pointing toward the opposite ASIS.

iii | **Movable arm:** Oriented along the anterior side of the femur.

iv | The line of the movable arm should be aligned with the 90° mark of the compass.

c | Directions

i | Without assistance, ask the patient to adduct the thigh as far as possible. ***Do not move the stationary arm.*** **Note:** Patient will most likely need to cross the moving limb over the stance leg.

ii | Move the movable arm to realign with the femur.

iii | Look for the line of the movable arm as it passes through the compass. Record the measurement in Table 9.4.

Table 9.4 | Adduction: 20° (normal measurement)

	1st Measurement	2nd Measurement	Average
Student 1			
Student 2			

5 | **Medial Rotation: 40° (normal measurement)**

a | Patient position

i | Seated on a table with the legs flexed and dangling.

b | Goniometer position

i | **Axis:** At the patella of the limb being tested.

ii | **Stationary arm:** Oriented down along the anterior tibia.

iii | **Movable arm:** Also aligned distally along the anterior tibia.

iv | The line of the moveable arm should be aligned with the 0° mark on the compass.

c | Directions

i | Ask the patient to medially rotate the femur as far as possible.

ii | While keeping the stationary arm in place, align the movable arm with the tibia and record the measurement. Record the measurement in Table 9.5.

Table 9.5 | Medial Rotation: 40° (normal measurement)

	1st Measurement	2nd Measurement	Average
Student 1			
Student 2			

6 | **Lateral Rotation: 50° (normal measurement)**

a | Patient position

i | Seated on a table with the legs flexed and dangling.

b | Goniometer position

i | **Axis:** At the patella of the limb being tested.

ii | **Stationary arm:** Oriented down along the anterior tibia.

iii | **Movable arm:** Also aligned distally along the anterior tibia.

iv | The line of the movable arm should be aligned with the 0° mark on the compass.

c | Directions

i | Ask the patient to laterally rotate the femur as far as possible.

ii | While keeping the stationary arm in place, align the movable arm with the tibia and record the measurement. Record the measurement in Table 9.6.

Table 9.6 | Lateral Rotation: 50° (normal measurement)

	1st Measurement	2nd Measurement	Average
Student 1			
Student 2			

Lab Activity 2 | Tibiofemoral Joint

The proximal bone in this joint in the femur and the distal bone is the tibia. This is a modified synovial, hinge joint that will allow movement primarily in the sagittal plane. There is a slight amount of rotation allowed at the joint, but that will not be assessed in this lab.

1 | **Flexion: 150° (normal measurement)**

a | Patient position

i | Standing and using a desk or a chair for support.

b | Goniometer position

i | **Axis:** Over the lateral condyle of the femur.

ii | **Stationary arm:** Oriented up along the lateral side of the femur.

iii | **Movable arm:** Aligned down along the lateral leg.

iv | The line of the movable arm should be aligned with the 0° mark on the compass.

c | Directions

i | Ask the patient to flex the leg as far as possible without any assistance. Make sure the patient is holding onto a desk or a chair for support. Ensure the stationary arm doesn't move.

ii | Realign the movable arm with the leg. Record the measurement in Table 9.7.

Table 9.7 | Flexion: 150° (normal measurement)

	1st Measurement	2nd Measurement	Average
Student 1			
Student 2			

Lab Activity 3 | Talocrural Joint

The proximal bone in this joint is the tibia with some involvement of the lateral malleolus. The distal bone is the talus. This is another synovial, hinge joint and as such there is only movement in the sagittal plane. There are two unique actions associated with this joint: dorsiflexion and plantar flexion.

1 | **Dorsiflexion: 30° (normal measurement)**

a | Patient position

i | Seated on a table with the legs flexed and dangling.

b | Goniometer position

i | **Axis:** The lateral malleolus of the fibula.

ii | **Stationary arm:** Aligned up along the lateral leg.

iii | **Movable arm:** Aligned with the long axis of the foot. At this point the foot needs to be at a 90° angle to the leg.

iv | The line of the moveable arm should be aligned with the 90° mark on the compass.

c | Directions

i | Ask the patient to bring the toes and the foot toward the ceiling as far as possible without any assistance. Ensure the stationary arm doesn't move.

ii | Realign the movable arm with the foot. Record the measurement in Table 9.8.

Table 9.8 | Dorsiflexion: 30° (normal measurement)

	1st Measurement	2nd Measurement	Average
Student 1			
Student 2			

2 | **Plantar Flexion: 40° (normal measurement)**

a | Patient position

i | Seated on a table with the legs flexed and dangling.

b | Goniometer position

i | **Axis:** The lateral malleolus of the fibula.

ii | **Stationary arm:** Aligned up along the lateral leg.

iii | **Moveable arm:** Aligned with the long axis of the foot. At this point the foot needs to be at a 90° angle to the leg.

iv | The line of the movable arm should be aligned with the 90° mark on the compass.

c | Directions

i | Ask the patient to point the toes and the foot toward the floor as far as possible without any assistance. Ensure the stationary arm doesn't move.

ii | Realign the movable arm with the foot. Record the measurement in Table 9.9.

Table 9.9 | Plantar Flexion: 40° (normal measurement)

	1st Measurement	2nd Measurement	Average
Student 1			
Student 2			

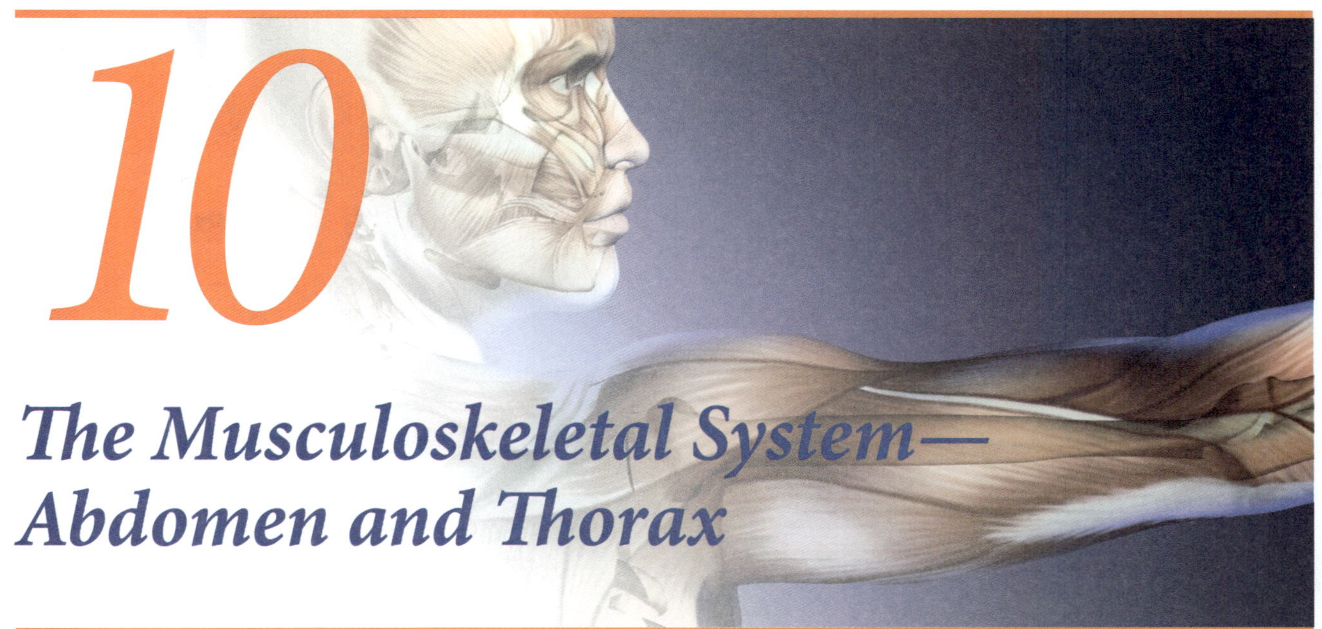

10 The Musculoskeletal System—Abdomen and Thorax

Lab Activity 1 | Bones of the Vertebral Column and Thoracic Cage

From the bone cases provided, take out the bones of the vertebral column and thoracic cage only. Using your textbook and other references as guides, identify the following bones and bone features.

A | Vertebral Column

1 | Features on all vertebrae

- Vertebral foramen
- Body
- Spinous process
- Transverse process
- Lamina
- Intervertebral foramen

2 | **Cervical vertebrae** (1st seven vertebrae)

- Transverse foramen
- Atlas
- Axis
- Dens (odontoid process)—found only on the axis

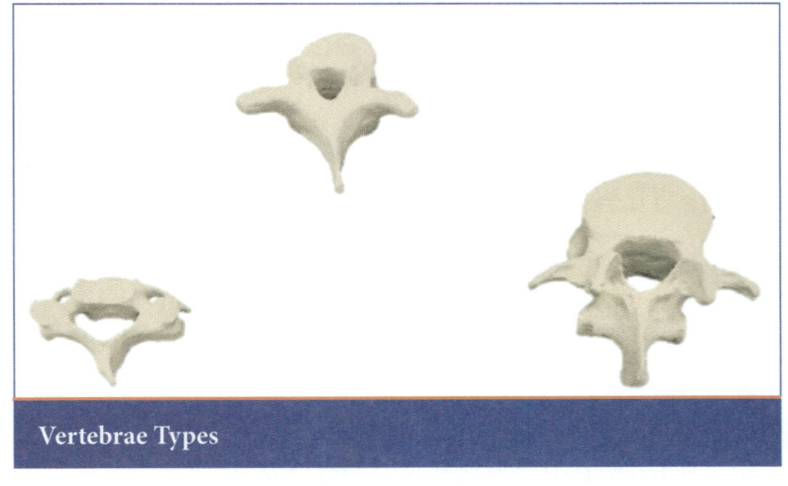

Vertebrae Types

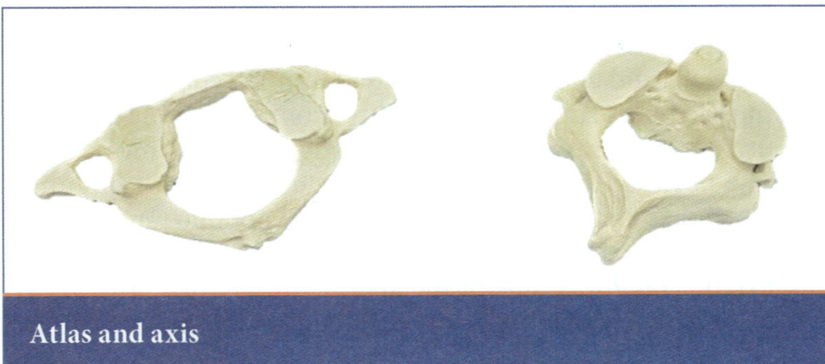

Atlas and axis

3 | **Thoracic vertebrae** (12 vertebrae)

4 | **Lumbar vertebrae** (5 vertebrae)

5 | **Sacrum** (5 fused)

6 | **Coccyx** (3–5 fused)

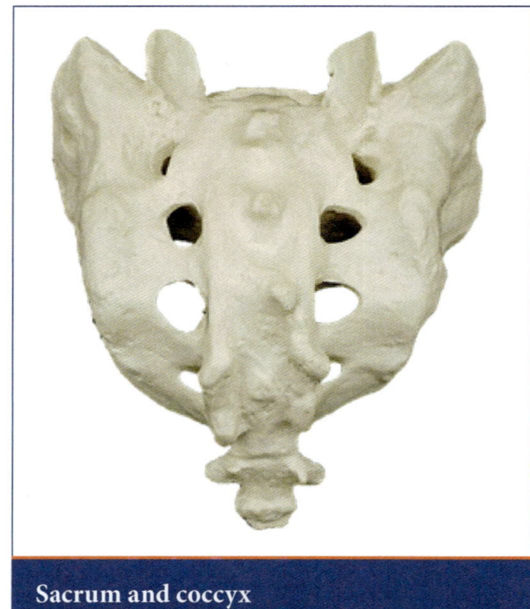

Sacrum and coccyx

B | Thoracic Cage

1 | **Ribs (costa)**

- True ribs (1st seven pairs)
- False ribs (ribs 8–10)
- Floating ribs (ribs 11 and 12)

2 | **Sternum**

- Manubrium
- Body
- Xiphoid process

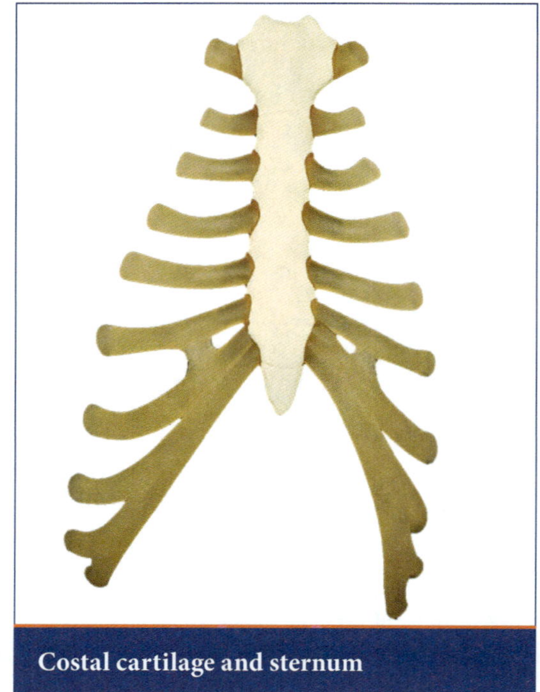

Costal cartilage and sternum

Lab Activity 2 | Costal Cartilage Muscles of the Thorax, Abdomen, and Back

Using your textbook and other reference materials as guides, locate the following muscles on the human cadavers and models. You will be responsible for the origins, insertions, and actions as well as identification.

Abdominal Muscles (Anterior)

Because the abdominal walls have no bony reinforcements, they rely on four sets of muscles to strengthen the walls and to help contain the abdominal contents. These muscles are layered on top of each other, and the muscle fibers run in different directions, providing strength in the abdominal wall. Plywood construction utilizes these same concepts (i.e., wood grains running in different directions) giving plywood its strength. The fibers of the external oblique muscle run at right angles to the fibers of the internal oblique muscle which lies just underneath the external oblique muscle. These muscles attach to aponeuroses (broad bands of connective tissue) and then to their insertions. The rectus abdominis muscle is a strap like muscle that extends on either side of the abdomen's midline. The insertions of the rectus abdominis muscle form the linea alba ("white line") which extends from the sternum to the pubic symphysis.

The abdominal muscles help to protect and support the viscera by compressing the abdomen. They work best when they are well toned. If they are not sufficiently toned or when they are severely stretched (as occurs during pregnancy or weight gain), they become weak and allow the abdomen to protrude (a "pot belly"). Additional functions of the abdominal muscles include lateral flexion of the vertebral column (envision "side bend" exercises), rotation of the vertebral column, and anterior flexion of the vertebral column against resistance (as in sit-ups). They are also involved in breathing movements, urination, defecation, vomiting, coughing, screaming, and childbirth.

Table 10.1 | Abdominal Muscles (Anterior)

Muscle Name	Origin (O) and Insertion (I)	Action
You need to be able to **identify** and **correctly spell** each muscle.	For each muscle, you **are required to** learn ***all*** of the origins and insertions listed below. Any items in parentheses ( ) may be optional based upon your individual instructor's preferences.	You **are required to** learn the action(s) indicated in **bold** type. The remainder of the material concerning actions is for informational purposes only.
Rectus abdominis	**O:** Pubic symphysis **I:** Xiphoid process and costal cartilages	**Compresses abdomen; flexes vertebral column;** retains contents of abdomen; stabilizes pelvis during walking.
External oblique	**O:** Outer surfaces of lower 8 ribs **I:** Linea alba	**Compresses abdomen; flexes vertebral column laterally.**
Internal oblique	**O:** Iliac crest **I:** Linea alba	**Compresses abdomen; flexes vertebral column laterally.**

Breathing Muscles

The primary muscles responsible for breathing are the deep muscles of the thorax. The intercostal muscles are short muscles extending from one rib to the next. When they contract, they pull the slightly flexible ribs closer together. The external intercostal muscles are superficial to the internal intercostal muscles. The most important muscle for breathing is the diaphragm, which forms a muscular divide between the thoracic and abdominal cavities. The contraction of all of these muscles causes pressure changes which are required for breathing. Forced expiration involves the internal intercostal muscles as well as some of the abdominal muscles.

Table 10.2 | Breathing Muscles

Muscle Name	Origin (O) and Insertion (I)	Action
You need to be able to **identify** and **correctly spell** each muscle.	For each muscle, you **are required to** learn ***all*** of the origins and insertions listed below. Any items in parentheses ( ) may be optional based upon your individual instructor's preferences.	You **are required to** learn the action(s) indicated in **bold** type. The remainder of the material concerning actions is for informational purposes only.
Diaphragm	**O:** Bottom of rib cage, sternum, and lumbar vertebrae **I:** Central tendon	Prime mover for **inspiration** (inhalation). The diaphragm flattens when it contracts and becomes dome-shaped when it relaxes.
External intercostals	**O:** Inferior border of rib above **I:** Superior border of rib below	Elevates the ribs for **inspiration.**
Internal intercostals	**O:** Superior border of rib below **I:** Inferior border of rib above	Depresses the ribs for **forced expiration.**

Posterior Muscles That Move the Vertebral Column

The deep muscles of the back are responsible for many trunk movements. (The superficial muscles of the back are involved primarily in shoulder and arm movements.) The numerous deep muscles of the back form a broad, thick column which extends from the skull to the sacrum. You might think of these muscles of varying lengths as strings which, when pulled, cause one or more vertebrae to be extended. The largest of the deep back muscles consists of a group of muscles called the erector spinae ("spine erectors" or "straighteners"). The origins and insertions of these muscles overlap so entire regions of the vertebral column can be moved simultaneously and smoothly. Other actions of the deep back muscles are extension of the spine and lateral flexion of the vertebral column.

Erector spinae: The erector spinae consist of three sets of muscles. These muscles are the prime movers for extension of the vertebral column (help return the body to upright position after bending forward). During full flexion of the vertebral column (such as touching fingers to toes), these muscles are relaxed. When you attempt to return to the upright position, these muscles are initially inactive, and extension is initiated by the hamstring muscles and gluteus maximus muscle. The erector spinae muscles are activated when lifting a heavy load or moving suddenly from a bent-over position which can cause injury to the muscles and ligaments of the back.

Table 10.3 | Posterior Muscles That Move the Vertebral Column

Muscle Name	Origin (O) and Insertion (I)	Action
You need to be able to **identify** and **correctly spell** each muscle.	For each muscle, you **are required to** learn ***all*** of the origins and insertions listed below.	You **are required to** learn the action(s) indicated in **bold** type. The remainder of the material concerning actions is for informational purposes only.
Erector Spinae Group		
Iliocostalis	**O:** Iliac crest **I:** Ribs	All of these muscles **extend the vertebral column** to help maintain our upright posture. They may also laterally flex the vertebral column to the same side.
Longissimus	**O:** Transverse processes of lumbar vertebrae **I:** Transverse processes of thoracic vertebrae	
Spinalis	**O:** Spinous processes of upper lumbar and lower thoracic vertebrae **I:** Spinous processes of upper thoracic vertebrae	

11

The Musculoskeletal System—Upper Extremity

Lab Activity 1 | Bones of the Shoulder and Arm

From the bone cases provided, take out the bones of the shoulder girdles, arms, and hands only. Using your textbook and other references as guides, identify the following bones and bone features.

A | Shoulder/Pectoral Girdle

1 | **Scapula**

- Glenoid cavity (fossa)
- Spine
- Acromion process
- Coracoid process
- Axillary (lateral) border
- Vertebral (medial) border
- Supraspinous fossa
- Infraspinous fossa

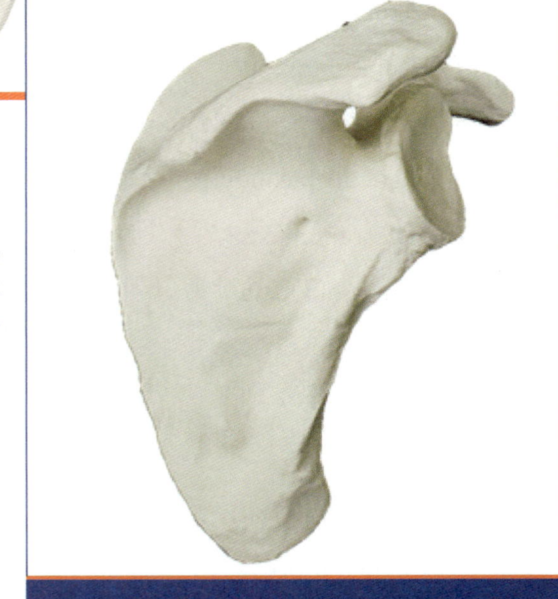

Scapula posterior

2 | **Clavicle**

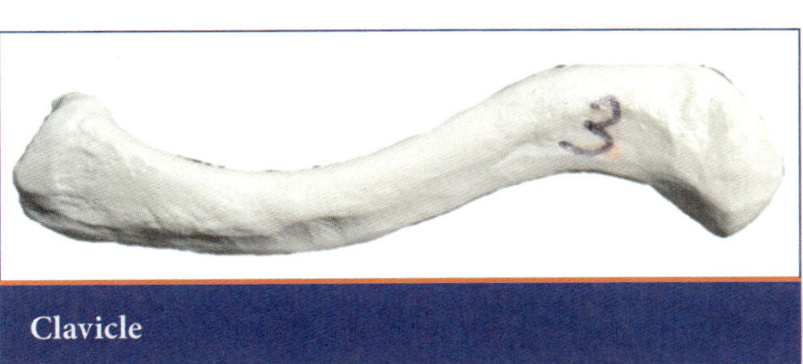

Clavicle

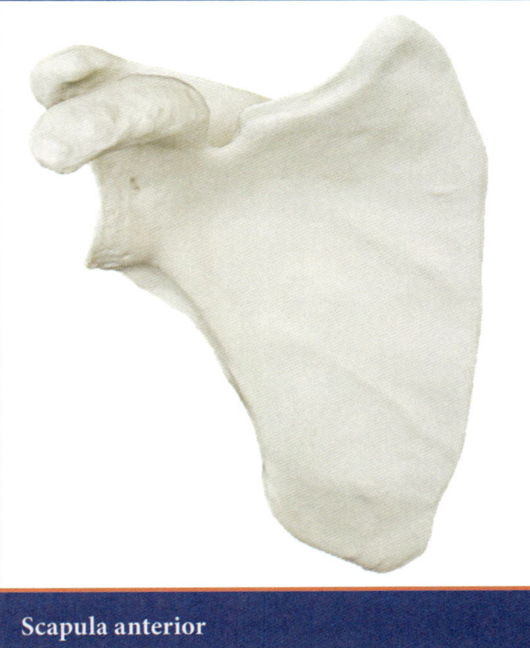

Scapula anterior

B | Upper Extremity

1 | Humerus

- Head
- Neck (surgical)
- Greater tubercle
- Lesser tubercle
- Intertubercular sulcus (intertubercular groove)
- Deltoid tuberosity
- Trochlea
- Capitulum
- Medial epicondyle
- Lateral epicondyle
- Olecranon fossa
- Coronoid fossa
- Radial fossa

2 | Radius

- Head
- Radial tuberosity
- Styloid process

3 | Ulna

- Trochlear notch
- Olecranon process
- Coronoid process

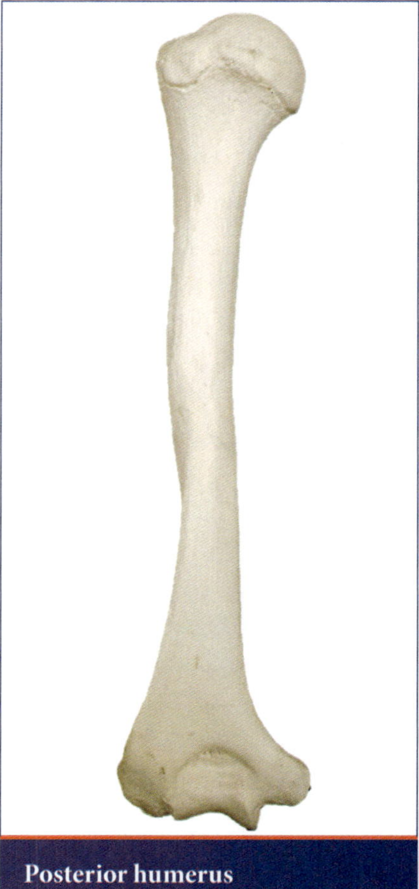

Posterior humerus

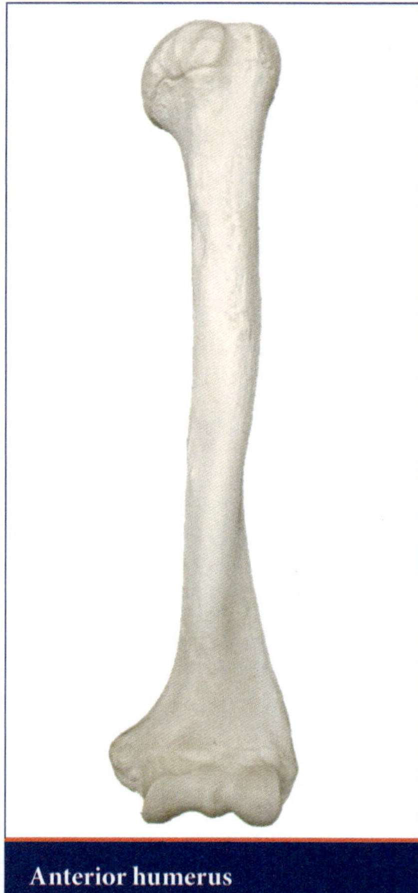

Anterior humerus

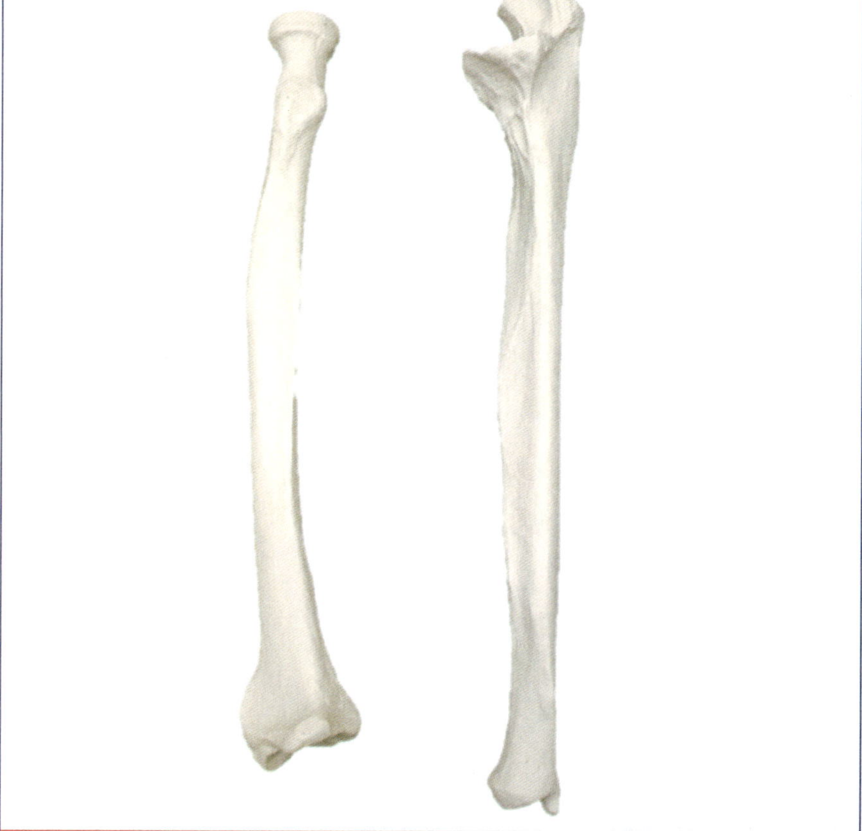

Radius and ulna

4 | **Carpals**

- Scaphoid
- Lunate
- Triquetrum (triquetral)
- Pisiform
- Trapezium
- Trapezoid
- Capitate
- Hamate

(Sally left the party to take Cathy home.)

5 | **Metacarpals 1–5 (must include the number)**

6 | **Phalanges 1–5 (must include the number)**

- Proximal phalanx
- Middle phalanx
- Distal phalanx

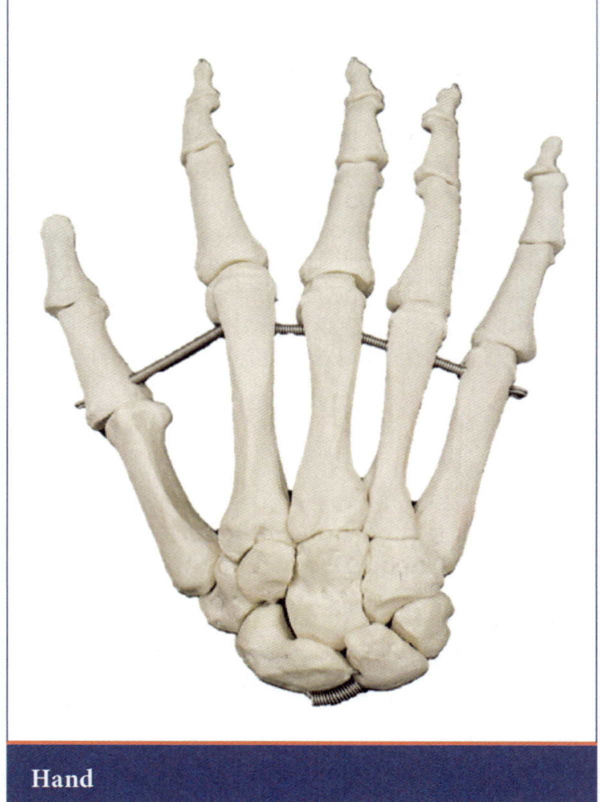

Hand

Muscles of the Shoulder and Arm

Muscles That Move the Scapula

Two of the muscles that move the scapula are located on the anterior side of the thorax—pectoralis minor and serratus anterior, and three are located on the posterior side of the thorax—trapezius, levator scapulae, and rhomboids. The important movements of the scapula are elevation, depression, rotation, abduction, and adduction. The clavicles help provide stability and precision to scapular movements. Due to the complexity of the shoulder structure, the attached muscles do not function independently. For example, to elevate the scapula, both the trapezius and levator scapulae muscles must work together.

Muscles That Move the Arm (Humerus)

Muscles that move the arm cross the shoulder joint. The shoulder joint is a ball-and-socket joint and is one of the most mobile joints in the body. Due to its mobility, it is unstable which leads to frequent injury. A total of nine muscles cross each shoulder joint and insert on the humerus (you will not be learning all nine). Of the nine muscles, the pectoralis major, latissimus dorsi, and deltoid muscles are prime movers. The supraspinatus, infraspinatus, teres minor, and subscapularis muscles are collectively known as rotator cuff muscles. They act as synergists in some of the arm movements, but their primary role is to hold the head of the humerus in the glenoid cavity (fossa).

Muscles That Move the Forearm (Radius and Ulna)

The muscles that move the forearm cross the elbow joint and insert on the forearm bones (radius and ulna). Because the elbow is a hinge joint, the primary movements that occur in the forearm are flexion and extension. The flexors are located anteriorly and the extensors are located posteriorly. There are also muscles in the forearm that act as pronators and supinators.

Muscles That Move the Wrist and Fingers

The muscles of the forearm are almost equally divided into muscles that cause movements of the wrist and muscles that cause movement of the fingers. Although some of the muscles originate on the humerus and can cross both the elbow and wrist joints, their actions on the elbow are so minimal that they are normally disregarded. The muscles contribute to the roundness seen in the forearm, but then the muscles taper to long insertion tendons. The insertions are anchored by strong ligaments known as retinacula ("retainers"). One of these retinacula (the flexor retinaculum) along with some of the carpals creates a space known as the carpal tunnel. The carpal tunnel is a passageway for the long flexor insertion tendons of the fingers and thumb and for the median nerve. When the carpal tunnel narrows, it gives rise to carpal tunnel syndrome, which is compression of the median nerve. The flexors are located on the anterior surface of the forearm and the extensors on the posterior surface.

Lab Activity

Using your textbook and other reference materials as guides, locate the following muscles on the human cadavers and models. You will be responsible for the origins, insertions, and actions as well as identification.

Table 11.1 | Anterior Muscles

Muscle Name	Origin (O) and Insertion (I)	Action
You need to be able to **identify** and **correctly spell** each muscle.	For each muscle, you **are required to** learn ***all*** of the origins and insertions listed below.	You **are required to** learn the action(s) indicated in **bold** type. The remainder of the material concerning actions is for informational purposes only.
	Muscles that Move the Scapula	
Pectoralis minor	**O:** Ribs 3–5 **I:** Coracoid process	**Depresses scapula.**
Serratus anterior	**O:** Ribs 1–9 **I:** Vertebral/medial border of scapula	**Abducts scapula.** This muscle is used in pushing and is important in horizontal arm movements. Sometimes referred to as the "boxer's" muscle.
	Muscles that Move the Humerus	
Pectoralis major	**O:** Sternum **I:** Greater tubercle	**Flexes and adducts humerus.**
Deltoid	**O:** Clavicle; acromion process and spine scapula **I:** Deltoid tuberosity	This is a large muscle whose role can change based upon which set of fibers (anterior or posterior) are contracting. **Abducts humerus** when all of its fibers contract at once; it is an antagonist of the pectoralis major and latissimus dorsi muscles. It is active during the arm swinging movements that occur with walking.

Muscles that Move the Forearm		
Biceps brachii	**O:** Coracoid process **I:** Radial tuberosity	**Flexes forearm.** Can also supinate forearm.
Brachialis	**O:** Anterior humerus **I:** Coronoid process	**Flexes forearm.** It contracts simultaneously with the biceps brachii muscle. The brachialis muscle pulls on the ulna as the biceps brachii muscle pulls on the radius.
Brachioradialis	**O:** Above lateral epicondyle of humerus **I:** Styloid process of radius	**Flexes forearm.** Synergist to biceps brachii muscle. Because the origin is at the distal humerus and the insertion at the distal forearm, this is a weak forearm flexor on its own. It becomes active when the elbow has been partially flexed by the prime movers (biceps brachii and brachialis muscles).
Muscles that Move the Wrist and Fingers Note: Please notice the position of the cadaver forearms. If the forearms are not in anatomical position, you will need to make adjustments as to the location of the wrist and finger flexors and extensors.		
Flexor carpi radialis	**O:** Medial epicondyle of humerus **I:** 2nd metacarpal The insertion tendon is easily seen and is used to guide for the position of the radial artery which is used for taking the pulse at the wrist.	**Flexes wrist.**
Palmaris longus (This muscles may or may not be present; it is absent in approximately 14% of the population.)	**O:** Medial epicondyle of humerus **I:** Palmar aponeurosis	**Flexes wrist.**
Flexor carpi ulnaris	**O:** Medial epicondyle of humerus **I:** 5th metacarpal	**Flexes wrist.**
Flexor digitorum	**O:** Medial epicondyle of humerus **I:** Phalanges 2–5	**Flexes phalanges/digits.**

Table 11.2 | Posterior Muscles

Muscle Name	Origin (O) and Insertion (I)	Action
You need to be able to **identify** and **correctly spell** each muscle.	For each muscle, you **are required to** learn ***all*** of the origins and insertions listed below.	You **are required to** learn the action(s) indicated in **bold** type. The remainder of the material concerning actions is for informational purposes only.

Muscles that Move the Scapula		
Trapezius	**O:** Occipital bone; spinous processes of C7 and thoracic vertebrae **I:** Clavicle; acromion process and spine of scapula	This is a large muscle whose role can change based upon which set of fibers (superior, middle, or inferior) are contracting. **Elevates scapula** (superior fibers); **Depresses scapula** (inferior fibers); **Adducts scapula** (middle fibers). The trapezius also stabilizes the scapula.
Levator scapulae	**O:** Transverse processes of cervical vertebrae **I:** Superior vertebral/medial border of scapula	**Elevates scapula** (works with trapezius). Also laterally flexes neck to same side.
Rhomboids	**O:** Spinous processes of upper thoracic vertebrae **I:** Vertebral/medial border of scapula	**Adducts scapula.**
Muscles that Move the Humerus		
Latissimus dorsi	**O:** Thoracolumbar fascia **I:** Intertubercular sulcus/groove	**Extends and adducts humerus.** This muscle is an antagonist to the pectoralis major. This muscle is well developed in swimmers.
Supraspinatus	**O:** Supraspinous fossa **I:** Greater tubercle	**Abducts humerus** along with the deltoid. It stabilizes the shoulder joint when performing tasks like carrying a heavy suitcase at your side.
Infraspinatus	**O:** Infraspinous fossa **I:** Greater tubercle	**Laterally rotates humerus;** helps hold head of humerus in glenoid cavity.
Teres major	**O:** Dorsal (posterior) surface of scapula **I:** Lesser tubercle	**Extends and adducts humerus** (synergist of latissimus dorsi muscle).
Teres minor	**O:** Axillary/lateral border of scapula **I:** Greater tubercle	**Laterally rotates humerus** (synergist of infraspinatus muscle).
Muscles that Move the Forearm		
Triceps brachii	**O:** Scapula and posterior humerus **I:** Olecranon process	**Extends forearm.**

Muscles that Move the Wrist and Fingers Note: Please notice the position of the cadaver forearms. If the forearms are not in anatomical position, you will need to make adjustments as to the location of the wrist and finger flexors and extensors.		
Extensor carpi radialis	**O:** Lateral epicondyle of humerus **I:** 2nd metacarpal	**Extends wrist.**
Extensor carpi ulnaris	**O:** Lateral epicondyle of humerus **I:** 5th metacarpal	**Extends wrist.**
Extensor digitorum	**O:** Lateral epicondyle of humerus **I:** Phalanges 2–5	**Extends phalanges/digits;** can abduct (flare) fingers.

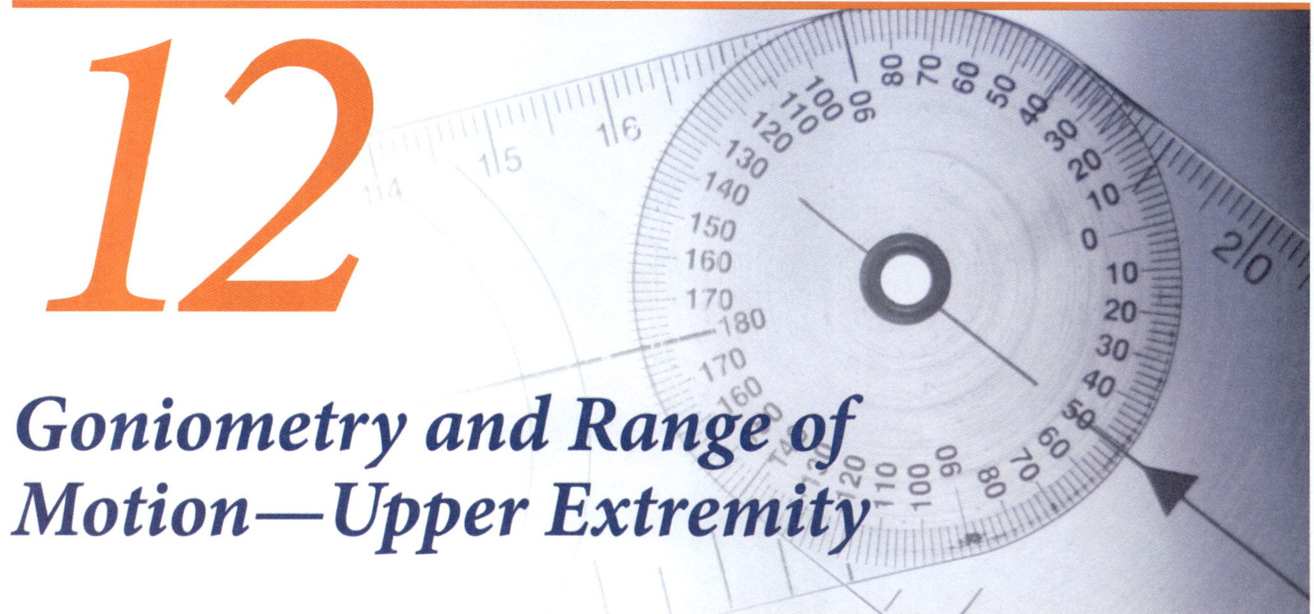

12

Goniometry and Range of Motion—Upper Extremity

Introduction

The purpose of this lab is to determine the active range of motion allowed at several different joints. A joint, or an articulation, is where two or more bones meet. Gross movements of the body occur at these joints and are caused by contractions of skeletal muscles. Assessing joint range of motion (ROM) is an important step in understanding the overall health of the joint.

Goniometers (gonio = angle, meter = device to measure) are used to assess ROM. When used properly, the measurements taken using a goniometer are both reliable and valid. Goniometry is used mainly in health and fitness facilities especially physical therapy, orthopedic, and chiropractic offices to help assess injured patients. Goniometers are assessment tools and can be used to obtain a baseline measurement and to determine the effectiveness of treatment. Each joint is unique in its structure and in the amount of movement allowed. The anatomical adage of form equals function is never so obvious as it is in the articulations of the skeleton.

The goniometer has three parts:

- The **axis** is the point where the two arms of the goniometer meet. The arms will be able to rotate around this axis. We will be using this axis to ensure proper positioning of the goniometer.
- The **stationary arm** of the goniometer has a compass with numbers arranged in a circle. These numbers are read as degrees. As the name implies, this arm will remain fixed when taking a measurement.
- The **movable arm** will have a line and/or arrow at its axial end that will rotate around the compass of the stationary arm. This arm will move with the patient's limb and the ROM can be assessed by reading the numbers in **black** (not red) through which the line passes.

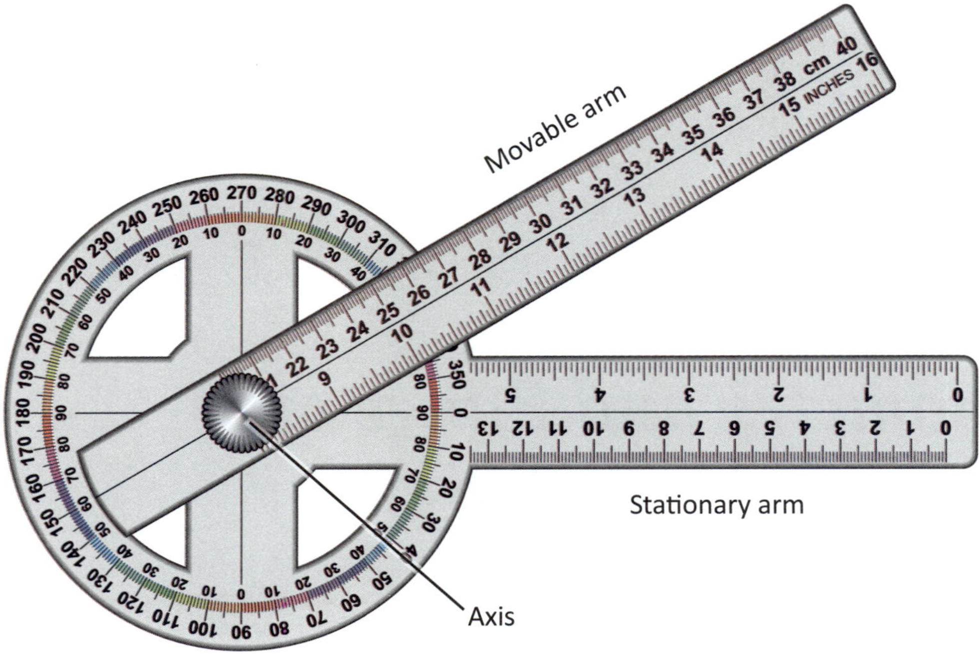

Figure 12.1 | Goniometer

Procedures

1 | Working with a lab partner, you will measure the ROM of each other's upper extremity joints.

2 | Each measurement will be taken two times, and the average calculated.

3 | When recording the measurement, use the **black numbers** on the goniometer.

4 | All data will be recorded in the tables.

Lab Note: In some clinical settings, goniometry measurements may be taken with the patient in a supine position.

Lab Activity 1 | Glenohumeral Joint

The proximal bone in this joint is the scapula with the joint surface being the glenoid fossa. The distal bone will be the humerus. Much like the hip, this joint will be a synovial, ball and socket, which will allow movement in all three planes of motion. This is the most mobile joint in the body, but this mobility comes at the price of stability. There is also considerable assistance to increase glenohumeral ROM from the scapulothoracic, acromioclavicular, and sternoclavicular joints.

1 | **Flexion: 180° (normal measurement)**

a | Patient position

i | Standing with the arms held loosely at the sides with the palms facing anteriorly.

b | Goniometer position

i | **Axis:** Over the greater tubercle.

ii | **Stationary arm:** Aligned down along the lateral side of the arm.

iii | **Movable arm:** Aligned down along the lateral side of the arm.

iv | The line of the movable arm should be aligned with the 0° mark on the compass.

c | Directions

i | Ask the patient to flex the arm as far as possible without assistance.

ii | Realign the movable arm with the stationary arm. Record the measurement in Table 12.1.

Table 12.1 | Flexion: 180° (normal measurement)

	1st Measurement	2nd Measurement	Average
Student 1			
Student 2			

2 | **Extension: 50° (normal measurement)**

a | Patient position

i | Standing with the arms held loosely at the sides with the palms facing anteriorly.

b | Goniometer position

i | **Axis:** Over the greater tubercle.

ii | **Stationary arm:** Aligned down along the lateral arm.

iii | **Movable arm:** Aligned down along the lateral arm.

iv | The line of the movable arm should be aligned with the 0° mark on the compass.

c | Directions

i | Ask the patient to extend the arm as far as possible without assistance.

ii | Realign the movable arm with the stationary arm. Record the measurement in Table 12.2.

Table 12.2 | Extension: 50° (normal measurement)

	1st Measurement	2nd Measurement	Average
Student 1			
Student 2			

3 | **Abduction: 180° (normal measurement)**

a | Patient position

i | Standing with the arms held loosely at the sides and the palms facing anteriorly.

b | Goniometer position

i | **Axis:** Over the lesser tubercle.

ii | **Stationary arm:** Aligned down along the anterior arm.

iii | **Movable arm:** Aligned down along the anterior arm.

iv | The line of the movable arm should be aligned with the 0° mark on the compass.

c | Directions

i | Ask the patient to abduct the arm as far as possible without assistance.

ii | Realign the movable arm with the stationary arm. Record the measurement in Table 12.3.

Table 12.3 | Abduction: 180° (normal measurement)

	1st Measurement	2nd Measurement	Average
Student 1			
Student 2			

4 | **Adduction: 50° (normal measurement)**

a | Patient position

i | Standing with the arms held loosely at the sides and the palms facing anteriorly.

b | Goniometer position

i | **Axis:** Over the lesser tubercle.

ii | **Stationary arm:** Aligned down along the anterior arm.

iii | **Movable arm:** Aligned down along the anterior arm.

iv | The line of the movable arm should be aligned with the 0° mark on the compass.

c | Directions

i | Ask the patient to adduct the arm as far as possible without assistance. The patient will most likely need to cross the arm in front of their torso to complete this movement.

ii | Realign the movable arm with the stationary arm. Record the measurement in Table 12.4.

Table 12.4 | Adduction: 50° (normal measurement)

	1st Measurement	2nd Measurement	Average
Student 1			
Student 2			

5 | **Lateral Rotation: 90° (normal measurement)**

a | Patient position

i | The patient will be standing.

ii | Ask the patient to flex the forearm to 90°.

iii | While keeping the forearm flexed, ask the patient to abduct the shoulder to 90°.

iv | At this point the palm of the hand should be facing the floor and the elbow joint should be pointing laterally.

b | Goniometer position

i | **Axis:** Over the olecranon process.

ii | **Stationary arm:** Aligned along the lateral forearm toward the fingers.

iii | **Movable arm:** Aligned along the lateral forearm toward the fingers.

iv | The line of the movable arm should be aligned with the 0° mark on the compass.

c | Directions

i | While keeping the forearm flexed to 90˚ and the shoulder abducted to 90°, ask the patient to raise the forearm as far as possible so the palm of the hand now faces anteriorly.

ii | Make sure that the patient's torso remains straight; the patient should not extend the torso. Only the arm should move.

iii | Realign the movable arm with the forearm. Record the measurement in Table 12.5.

Table 12.5 | Lateral Rotation: 90° (normal measurement)

	1st Measurement	2nd Measurement	Average
Student 1			
Student 2			

6 | **Medial Rotation: 90° (normal measurement)**

a | Patient position

i | The patient will be standing.

ii | Ask the patient to flex the forearm to 90°.

iii | While keeping the forearm flexed, ask the patient to abduct the shoulder to 90°.

iv | At this point the palm of the hand should be facing the floor and the elbow joint should be pointing laterally.

b | Goniometer position

i | **Axis:** Over the olecranon process.

ii | **Stationary arm:** Aligned along the lateral forearm toward the fingers.

iii | **Movable arm:** Aligned along the lateral forearm toward the fingers.

iv | The line of the movable arm should be aligned with the 0° mark on the compass.

c | Directions

i | While keeping the forearm flexed to 90° and the shoulder abducted to 90°, ask the patient to lower the forearm as far as possible so the palm of the hand now faces posteriorly.

ii | Realign the movable arm with the forearm. Record the measurement in Table 12.6.

Table 12.6 | Medial Rotation: 90° (normal measurement)

	1st Measurement	2nd Measurement	Average
Student 1			
Student 2			

Lab Activity 2 | Elbow Joint

The proximal bone in the joint is the humerus. It will articulate at the elbow with both the radius and the ulna. It is important to note that the proximal radius and ulna form a separate joint called the proximal radioulnar joint. It is at this joint, along with the distal radioulnar joint, that pronation and supination occur. The elbow is considered a synovial hinge joint and will only allow motion in the sagittal plane.

1 | **Flexion: 140° (normal measurement)**

- a | Patient position
  - i | The patient will be standing with the arms loosely at the side with the palm facing anteriorly.
- b | Goniometer position
  - i | **Axis:** Over the lateral epicondyle of the humerus.
  - ii | **Stationary arm:** Oriented up along the outside of the humerus.
  - iii | **Movable arm:** Oriented down along the outside of the forearm.
  - iv | The line of the movable arm should be aligned with the 0° mark on the compass.
- c | Directions
  - i | Ask the patient to flex the elbow as far as possible without any assistance.
  - ii | Realign the movable arm with the forearm. Record the measurement in Table 12.7.

Table 12.7 | Flexion: 140° (normal measurement)

	1st Measurement	2nd Measurement	Average
Student 1			
Student 2			

Lab Activity 3 | Wrist Joint

The proximal bone in the wrist is the radius. The distal bones of the joint are the scaphoid and lunate. The ulna does not articulate with any of the carpal bones and does not help form the wrist joint. This joint is a synovial condyloid joint and allows movement in two different planes. We will be assessing only flexion and extension, but the wrist also allows radial and ulnar deviation.

1 | **Flexion: 60° (normal measurement)**

- a | Patient position
  - i | The patient will be standing with the elbow flexed at 90°.
  - ii | The palm of the hand should be facing superiorly with the fingers extended.
- b | Goniometer position
  - i | **Axis:** Placed over the styloid process of the radius.
  - ii | **Stationary arm:** Oriented proximally along the lateral side of the forearm.
  - iii | **Movable arm:** Oriented distally along the outside of the hand.
  - iv | The line of the movable arm should be aligned with the 0° mark on the compass.

c | Directions

i | While keeping the fingers extended, ask the patient to flex the wrist as far as possible without assistance.

ii | Realign the movable arm with the hand. Record the measurement in Table 12.8.

Table 12.8 | Flexion: 60° (normal measurement)

	1st Measurement	2nd Measurement	Average
Student 1			
Student 2			

2 | **Extension: 60° (normal measurement)**

a | Patient position

i | The patient will be standing with the elbow flexed at 90°.

ii | The palm of the hand should be facing superiorly with the fingers extended.

b | Goniometer position

i | **Axis:** Placed over the styloid process of the radius.

ii | **Stationary arm:** Oriented proximally along the lateral side of the forearm.

iii | **Movable arm:** Oriented distally along the outside of the hand.

iv | The line of the movable arm should be aligned with the 0° mark on the compass.

c | Directions

i | While keeping the fingers extended, ask the patient to extend the wrist as far as possible without assistance.

ii | Realign the movable arm with the hand. Record the measurement in Table 12.9.

Table 12.9 | Extension: 60° (normal measurement)

	1st Measurement	2nd Measurement	Average
Student 1			
Student 2			

13

The Musculoskeletal System—Head and Neck

Lab Activity 1 | Facial and Cranial Bones

From the bone cases provided, take out the complete skull and the hyoid bone. Using your textbook and other references as guides, identify the following bones and bone features.

A | Cranial Bones and Sutures

1 | **Frontal bone**

2 | **Parietal bone**

3 | **Sagittal suture**

4 | **Coronal suture**

5 | **Lambdoid (lambdoidal) suture**

6 | **Squamous (squamosal) suture**

7 | **Sphenoid bone** *(You will not need to identify this bone, but you will need to know the parts listed below.)*

- Optic foramen (canal) *(for passage of optic nerve)*
- Hypophyseal fossa (sella turcica)

8 | **Temporal bone**

- Mastoid process
- Styloid process
- External auditory meatus (canal)
- Jugular foramen *(for passage of jugular vein, cranial nerves IX, X, XI)*
- Carotid foramen (canal) *(for passage of carotid artery)*
- Mandibular fossa
- Temporomandibular joint (TMJ)
- Zygomatic arch

9 | **Occipital bone**

- Foramen magnum
- Occipital condyle

10 | **Ethmoid bone** *(You will not need to identify this bone, but you will need to know the parts listed below.)*

- Middle nasal concha
- Crista galli *(attachment of meninges)*
- Cribriform plate *(passage of olfactory nerve)*

B | Facial Bones

1 | **Maxilla**

- Palatine process
- Alveolar process

2 | **Lacrimal bones**

3 | **Nasal bones**

4 | **Palatine bones**

5 | **Inferior nasal concha**

6 | **Vomer**

7 | **Zygomatic bones**

- Zygomatic arch

8 | **Mandible**

- Mandibular condyle (condylar process)
- Alveolar process

C | Hyoid Bone

Lab Activity 2 | Muscles of the Face, Head, and Neck

Using your textbook and other reference materials as guides, locate the following muscles on the head models. You will be responsible for the identification and actions ***only*** for this set of muscles.

Muscles of Facial Expression and Mastication

Superficial scalp and face muscles are involved in important facial expression movements. These muscles are highly variable in shape and in strength. They are also unusual in that they may not insert into bones but into other muscles or the skin.

Table 13.1 | Muscles of Facial Expression and Mastication

Muscle Name	Action	Origin (O) and Insertion (I)
You need to be able to **identify** each muscle.	For each muscle in this section, you may **choose one action** to learn for the practical.	You will not need to learn the origins and insertions for the practical on the muscles of the face and neck.
Frontalis	Raises eyebrows (as in surprise); wrinkles forehead horizontally.	**O:** Galea aponeurotica (on top of head) **I:** Skin of eyebrows and root of nose
Orbicularis oculi	Closes eye; blinking, squinting; draws eyebrows downward (which can produce "crow's feet" with aging).	**O:** Frontal and maxillary bones; ligaments around eyes **I:** Tissues of eyelids
Orbicularis oris	Shapes lips for speech and kissing; closes lips.	**O:** Maxilla and mandible **I:** Muscles and skin around angles of mouth
Zygomaticus major	Raises corners of mouth upward for smiling.	**O:** Zygomatic bone **I:** Skin and muscle at corner of mouth
Buccinator	Blowing; sucking (well developed in nursing infants); holds food between teeth when chewing.	**O:** Molar region of maxilla and mandible **I:** Orbicularis oris muscle
Masseter	Elevates mandible.	**O:** Zygomatic bone **I:** Mandible
Temporalis	Elevates mandible.	**O:** Temporal bone **I:** Top of mandible
Digastric	Depresses mandible.	**O:** Lower edge of mandible; mastoid process **I:** Hyoid bone
Stylohyoid	Elevates hyoid bone for swallowing.	**O:** Styloid process **I:** Hyoid bone

Muscles That Move the Head at the Neck

Many head movements occur through the actions of muscles that originate on the axial skeleton.

Table 13.2 | Muscles of the Head at the Neck

Muscle Name	Action	Origin (O) and Insertion (I)
You need to be able to **identify** each muscle.		You will not need to learn the origins and insertions for the practical on the muscles of the face and neck.
Sternocleidomastoid	Prime mover for neck flexion (bending the head as if praying).	**O:** Manubrium and clavicle **I:** Mastoid process
Splenius capitis	Extends neck; rotates head towards the same side.	**O:** Spinous processes of cervical and thoracic vertebrae **I:** Mastoid process; transverse processes of upper cervical vertebrae

14

Brain Anatomy and Cranial Nerves

Lab Activity

Identify the following structures using the reference materials provided, your text, models, and preserved brains.

A | Sagittal View

1 | **Brainstem:** The brainstem extends as a continuation of the spinal cord. It consists of three parts:

a | **Medulla:** Contains areas for control of the respiratory and cardiovascular systems. It is also where the last five cranial nerves originate: vestibulocochlear (VIII), glossopharyngeal (IX), vagus (X), accessory (XI), and hypoglossal (XII).

b | **Pons:** A bridge extending from the medulla to the structures above it. It is also where the following four cranial nerves originate: trigeminal (V), abducens (VI), facial (VII), and vestibulocochlear (VIII).

c | **Midbrain:** It is located superior to the pons. The midbrain is the origin of the oculomotor (III) and trochlear (IV) nerves.

2 | **Diencephalon:** Located superior to the brain stem. It contains three structures:

a | **Hypothalamus:** Has many functions including control of the endocrine system (hormones) and autonomic nervous systems. If present, the pituitary gland is attached to the hypothalamus by a stalk known as the infundibulum.

b | **Thalamus:** Oval structures superior to the hypothalamus. The thalamus functions as a route for sensory information.

c | **Pineal gland:** Located near the hypothalamus, this gland secretes the hormone ***melatonin.*** In the dissected sheep brains, this is often missing.

3 | **Ventricles of the brain:** The ventricles are cavities which contain cerebrospinal fluid. This fluid flows through the ventricles, the subarachnoid space, and the central canal. There are four ventricles:

a | **Lateral ventricles** (or ventricles 1 and 2) are found in each cerebral hemisphere. They are separated by a thin membrane called the ***septum pellucidum.***

b | The **third ventricle** is located in the space between the two masses of the thalamus. Connecting the third ventricle to the fourth ventricle is the thin canal known as the ***cerebral aqueduct.*** You can locate this structure as it passes through the midbrain.

c | The **fourth ventricle** is located between the pons and cerebellum. This ventricle connects with the ***central canal*** of the spinal cord.

4 | **Corpus callosum:** Near the top of the lateral ventricles is a thick, comma-shaped band of white tissue known as the ***corpus callosum.*** This structure is composed of commissural nerve fibers, which connect the right and left cerebral hemispheres and allow the two halves of the cerebrum to communicate.

5 | **Fornix:** Located near the bottom of the lateral ventricles is the ***fornix,*** which is part of the diencephalon and also composed of white fibers.

6 | **Cerebellum:** This is the second largest area of the brain and it has two separate hemispheres. The interior branching of white matter is known as the ***arbor vitae*** which resembles a piece of cut cauliflower.

7 | **Cerebrum:** The largest part of the brain, the cerebrum is divided into lobes ***(frontal, parietal, temporal, occipital)*** corresponding to the cranial bones which cover them.

B | *Superior View*

1 | **Fissures:** Are deep grooves in the brain.

a | **Longitudinal fissure** separates the two cerebral hemispheres (can only be viewed on an intact brain).

b | **Transverse fissure** separates the cerebrum from the cerebellum.

2 | **Gyri** (singular is gyrus) are the ridges of the cerebrum ("noodles"). Most have specific functions.

3 | **Sulci** (singular is sulcus) are the shallow grooves between the gyri of the cerebrum. Several sulci serve as landmarks to separate the cerebral lobes (frontal, parietal, temporal and occipital).

C | *Inferior View*

1 | **Cranial nerves:** There are twelve pairs of cranial nerves. Each pair has a name and is assigned a Roman numeral which represents the order the nerves are positioned along the brain from most anterior (I) to most posterior (XII).

Because of their delicate nature, it will be difficult to view many of these cranial nerves on the preserved specimens. You will need to identify the following structures:

Cranial Nerves

- Olfactory nerves
- Optic nerves
- Oculomotor nerves

Tracts

- Olfactory tracts
- Optic tracts

2 | There is an X-shaped structure known as the ***optic chiasm (chiasma)*** associated with the hypothalamus. The ***optic nerves*** from the eyes are the anterior legs extending from the middle of the X (they are often cut ***very*** short), and the optic tracts are the posterior legs of the X. The ***optic tracts*** transmit nerve impulses for vision to the visual cortex of the occipital lobes.

Required Structures of the Brain

You will be responsible for identifying and correctly spelling the following structures for a lab practical on the sheep brain and human brain, if available.

- Lobes—frontal, parietal, temporal, occipital
- Cerebrum
- Cerebellum
- Longitudinal fissure
- Transverse fissure
- Gyrus
- Sulcus
- Central canal
- Olfactory bulb
- Olfactory tract
- Optic nerve
- Optic chiasm (chiasma)
- Optic tract
- Oculomotor nerve
- Midbrain
- Pons
- Medulla
- Spinal cord
- Corpus callosum
- Fornix
- Thalamus
- Hypothalamus
- Pineal gland (body) (often missing)
- Cerebral aqueduct
- Arbor vitae
- Ventricles 1, 2, 3, 4 (1 and 2 also called lateral)

Laboratory 14 Review Questions

Brain Anatomy and Cranial Nerves

Name: ______________________________

Match each of the following structures with their functions.

1 | __________ Shallow groove

2 | __________ Contains centers for control of respiration and heart rate

3 | __________ Deep groove

4 | __________ Left or right half of cerebrum

5 | __________ Connects cerebral hemispheres

6 | __________ Secretes melatonin

7 | __________ Intersection of optic nerves and optic tracts

8 | __________ Folding of cerebral gray matter

9 | __________ Connects 3rd and 4th ventricles

10 | __________ Relay station for sensory impulses

11 | __________ Chambers for cerebrospinal fluid flow

12 | __________ Used in balance and equilibrium

a | Cerebellum

b | Fissure

c | Hemisphere

d | Optic chiasm(a)

e | Cerebral aqueduct

f | Sulcus

g | Gyrus

h | Medulla

i | Ventricles

j | Thalamus

k | Pineal gland (body)

l | Corpus callosum

13 | List the cranial nerves which arise from the following:

a | Medulla ______________________________

b | Pons ______________________________

c | Midbrain ______________________________

15

Neural Testing Sensory and Motor Functions of the Spinal Nerves

Identify the following structures using the reference materials provided, your text, and models.

Introduction to the Spinal Cord

1 | There are 31 segments of the spinal cord. From each segment are 31 pairs of spinal nerves. Axons from each segment synapse with those of adjacent segments to form peripheral nerves that receive sensory information from and provide motor commands to the various parts of our body.

2 | Thoracic, lumbar, and sacral spinal nerves are named according to the vertebral segment above. Cervical spinal nerves are named according to the vertebral segment below. Please see Table 15.1.

Table 15.1 | Cervical Spinal Nerves

Spinal Nerve	Number of Pairs	Characteristics
Cervical	8	C1 nerve lies between occipital bone and atlas. C2–7 nerves lie above the vertebrae with the same number. C8 lies between the C7 and T1 vertebrae.
Thoracic	12	T1–T12 nerves lie below the vertebrae with the same number (named for the vertebra above the nerve).
Lumbar	5	L1–L5 nerves lie below the vertebrae with the same number (named for the vertebra above the nerve).
Sacral	5	The sacrum was composed of five separate bony segments that fused together. The S1–S5 nerves are named for the vertebral segment above the nerve.
Coccygeal	1	Abbreviated Co.

3 | Each spinal nerve is formed from the union of a **dorsal nerve root** and a **ventral nerve root.** The dorsal nerve root is the sensory portion of the nerve, receiving input from the external and internal environments. The ventral nerve root is the motor portion of the nerve, sending commands to skeletal muscle, organs, and glands. The dorsal and ventral nerve roots at each segment merge to form a **mixed** spinal nerve. The term "mixed" is used to indicate that the nerve contains both sensory and motor fibers. All 31 pairs of spinal nerves are mixed.

4 | Each pair of spinal nerves monitors a specific region of the body. The area of skin supplied by a single spinal nerve is called a **dermatome.** Dermatome maps are often used clinically in the diagnostic process. Loss of skin sensation or altered sensation in a particular dermatome may indicate injury to a particular spinal nerve. A dermatome map is included in this lab.

5 | The spinal cord is composed of gray and white matter. The white matter is organized into **columns,** or *funiculi,* which are composed of **tracts,** or *fasciculi.* Tracts are organized into two groups of axons within the white matter of the spinal cord, sensory and motor. Sensory tracts carry action potentials along **ascending** pathways from peripheral receptors to the brain. Motor tracts carry action potentials from the brain to effectors along **descending** pathways. Different tracts within different pathways carry different types of information. Please see Table 15.2 for a list of the major tracts and pathways in the CNS.

Table 15.2 | Major Tracts and Pathways in the CNS

Pathway	Type of Information Carried
Posterior column pathway	Fine touch, pressure, vibration, proprioception
Anterior spinothalamic tract	Crude touch, pressure
Lateral spinothalamic tract	Pain, temperature
Spinocerebellar pathway	Unconscious proprioception
Corticospinal pathway	Motor commands to skeletal muscle

Materials Needed to Complete This Lab

- Reflex hammer
- Two-point discriminator

Lab Activity 1 | Deep Tendon Reflexes

Deep Tendon Reflexes: Deep tendon reflexes (DTRs), also referred to as *stretch reflexes,* are commonly utilized to assess the sensory and motor functions of a spinal nerve. A reflex is an involuntary, automatic response to a stimulus that does not require the brain for conscious thought. A reflex must always begin with a stimulus which activates a receptor. Here, the receptor is called a **muscle spindle.** Once activated, the muscle spindles carry afferent signals along the sensory neuron to the spinal cord, where a single synapse occurs with a motor neuron, which carries the command to the effector. In this case, the effectors are skeletal muscle.

There are several common areas to assess DTRs, and each corresponds to a particular spinal nerve. See Table 15.3.

Table 15.3 | Deep Tendon Reflexes (DTRs)

Reflex	Spinal Nerve(s)
Biceps brachii	C5/C6
Brachioradialis	C6
Triceps brachii	C7
Patellar (knee-jerk)	L3/L4
Achilles (ankle-jerk)	S1

It is important to understand how to evaluate one's reflex response. Table 15.4 demonstrates the scale commonly used to grade reflexes. Your instructor will describe and demonstrate what each may look like, and instructions for performing each are provided. Keep in mind that as you perform reflex testing, the area where you "tap" the hammer is the insertion point of that muscle. That means that the "jerk" response should be the action of that muscle.

Table 15.4 | Scale Commonly Used to Grade Reflexes

Response	Grade	Interpretation
Reflex absent (with reinforcement)	0/5	Reflex absent with reinforcement (see below)
Reflex with reinforcement	+1/5	No reflex on initial attempt. To *reinforce,* ask the patient to either clasp their hands (for lower extremity reflexes) or squeeze their knees together for upper extremity reflexes). While performing this, check the reflex again and document +1/5 if you observe a jerk.
Normal	+2/5	Average, expected response
Brisk	+3/5	More rapid than normal
Intermittent clonus	+4/5	Two "jerks" of the limb, abnormal response
Sustained clonus	+5/5	More than two "jerks" of the limb, abnormal response

For this activity, you will use a reflex hammer to assess each of the deep tendon reflexes listed in Table 15.3 and grade them according to Table 15.4. Perform on both sides of the body, document the findings in Table 15.5, and then evaluate the results. For each test, follow the instructions below:

1 | **Biceps brachii**

a | Stand in front of your lab partner who should be seated.

b | Ask your partner to keep their arm relaxed and positioned supine.

c | Locate the biceps brachii insertion tendon (your partner may need to flex their arm for you to locate it).

d | Place your thumb on the tendon, and while gently holding their relaxed arm for support, tap your thumb with the reflex hammer.

e | You should see contraction of the biceps brachii muscle and flexion of the forearm.

f | Record your results in Table 15.5.

2 | **Brachioradialis**

a | Stand in front of your lab partner who should be seated.

b | Ask your partner to keep their right arm relaxed and positioned supine. Their arm should be resting on their lap.

c | Locate the styloid process of the radius, which will feel like a bony bump near the thumb side of the wrist, and move proximally about 2 inches.

d | Place your thumb on the brachioradialis insertion tendon and tap your thumb with the reflex hammer.

e | You should see contraction of the brachioradialis muscle and slight flexion of the forearm.

f | Record your results in Table 15.5.

3 | **Triceps brachii**

a | Standing behind your lab partner, hold their relaxed limb so their arm is horizontal and their forearm is relaxed and hanging down.

b | Locate the triceps brachii insertion tendon (your partner may need to extend their arm for you to locate it).

c | Place your thumb on the tendon, and while gently holding their relaxed arm for support, tap the tendon with the reflex hammer.

d | You should see contraction of the triceps brachii muscle and extension of the forearm.

e | Record your results in Table 15.5.

4 | **Patellar**

a | Stand in front of your lab partner who should be seated high enough so their legs are relaxed and do not reach the floor.

b | Locate the patellar tendon, which is the insertion of the quadriceps femoris group, just below the patella but above the tibial tuberosity (your partner may need to extend their leg for you to locate it).

c | Tap the tendon with the reflex hammer.

d | You should see contraction of the quadriceps femoris muscles and extension of the leg.

e | Record your results in Table 15.5.

5 | **Achilles**

a | Kneel in front of your lab partner who should be seated high enough so their legs are relaxed and do not reach the floor.

b | Dorsiflex their foot and locate the Achilles tendon (depending on the type of shoes worn, you may need to ask your partner to take their shoes off).

c | While dorsiflexing their foot, tap the Achilles tendon with the reflex hammer.

d | You should see contraction of the calf muscles and plantar flexion.

e | Record your results in Table 15.5.

Table 15.5 | Results Table: Deep Tendon Reflexes

Reflex	Grade (R)	Grade (L)	If any responses are abnormal, what may account for this?
Biceps brachii			
Brachioradialis			
Triceps brachii			
Patellar (knee-jerk)			
Achille's (ankle-jerk)			

What may cause an abnormal DTR?

- Improper technique
- Damage to the spinal cord, nerve root, or peripheral nerve
- Damage to the muscle
- Disruption to the sensory and/or motor fibers

Lab Activity 2 | Two-Point Discrimination Test

This test is commonly used to assess a patient's ability to identify two close points on an area of skin. The density of touch receptors in our skin varies. In areas with a higher density of touch receptors, the person will be better able to discriminate between the two points compared to areas with a lower density of touch receptors.

In this activity you will evaluate your partner's ability to feel one or two points of contact on various parts of their skin. The table provided lists twelve areas of skin that will be tested.

1 | Begin by **predicting** which area you believe has the highest density of touch receptors compared to the lowest, and order them 1–12 (1 = highest density, 12 = lowest density). Put your predictions in Table 15.6.

2 | Next, you will use a 2-point discriminator to evaluate your partner. You will use the 5 mm setting on the 2-point discriminator.

3 | Your partner's eyes should be closed. Randomly apply one or both points of the 5 mm measurement on each region of the skin listed in Table 15.6. Ask your partner if they feel one or two points when you touch their skin with the 2-point discriminator. ***Only record results when you are touching your partner with two points.***

4 | Document the findings of each area, and repeat on the other side of the body.

5 | **Using provided alcohol wipes, disinfect the 2-point discriminator between each person.**

Table 15.6 | Results Table: Two-Point Discrimination Test

Area of Skin	Predict the area of highest density of touch receptors (1) to lowest density (12).	Response to touch with calipers—document "1" or "2" points detected. If nothing detected, write "0" (R).	Response to touch with calipers—document "1" or "2" points detected. If nothing detected, write "0" (L) .
Tip of 2nd finger			
Tip of thumb			
Center of palm			
Back of hand			
Anterior medial forearm			
Anterior lateral forearm			
Posterior medial forearm			
Posterior lateral forearm			
Anterior arm			
Posterior arm			
Top of shoulder			
Back of shoulder			

Answer the following questions concerning two-point discrimination tests.

1 | Did your results match your prediction? Explain why or why not.

2 | Were there any differences between the right and left side? Explain.

3 | What do the findings indicate about the density of touch receptors in different areas of the skin?

4 | Which spinal pathway/tract does this test evaluate?

5 | For any areas where no sensation was detected, which dermatome(s) may be affected? Use the provided dermatome map to help you answer this question.

Lab Activity 3 | Romberg Test

The **Romberg Test** is often used to assess ***proprioception.*** Proprioception refers to your body's ability to sense its movement and position. One example of this is your ability to touch your nose with your eyes closed. We utilize various types of proprioceptors, including ***golgi tendon organs*** at the interface of a muscle belly and its tendon; ***muscle spindles*** in skeletal muscle; ***mechanoreceptors*** in joint capsules; and the vestibular system in the ears.

Procedure

1 | Your lab partner will stand upright, keeping their feet close together, and their arms outstretched in front of them.

2 | Now, ask your partner to close their eyes and watch them carefully!

3 | Excessive swaying when their eyes are closed indicates a positive Romberg test.

4 | Document your findings below.

Romberg Test Results

☐ Normal Response ☐ Abnormal Response

What spinal pathway/tract does this test evaluate? ______________________________

____

Lab Activity 4 | Pathological Reflexes

Pathological reflexes are described as reversions to primitive responses as a result of damage to upper motor neurons in descending tracts, leading to loss of cortical inhibition or an overreaction to stimuli. They are often performed to confirm normal function of the central nervous system or determine if an underlying neurological condition may be present.

In this lab activity we will perform two commonly utilized pathological reflexes: the Babinski reflex for the lower extremity and the Hoffman reflex for the upper extremity.

1 | **Babinski Reflex**

a | Ask for volunteers to remove their shoes and socks.

b | Clean the reflex hammer with the provided alcohol wipes prior to performing the Babinski test.

c | Using the metal end of a reflex hammer, scrape the bottom of the foot in an upside down "J" pattern, starting at the lateral side of the heel and ending medially near the hallux.

d | A normal response is rapid plantar flexion of the hallux.

e | Slower extension (dorsiflexion) of the hallux with fanning of the other toes indicates an abnormal response, or a positive Babinski sign.

f | Repeat on the other side and document your findings in Table 15.7.

g | Please note that positive findings seen in an adult (hallux extension) are considered normal in newborns up to about age 2.

h | After you have finished testing, clean the reflex hammer with alcohol wipes.

2 | **Hoffman Reflex**

a | Take your lab partner's relaxed hand and position the hand so it is prone with slight wrist extension and flexion of the fingers.

b | Hold their middle finger with your index and middle fingers at the proximal interphalangeal joint (second knuckle).

c | Using your thumb, sharply flick the nail of your partner's middle finger.

d | Flexion of the middle finger followed by relaxation indicates a normal response.

e | Flexion and adduction of the thumb with flexion of the index finger indicates an abnormal response, or a positive Hoffman sign.

f | Repeat on the other side and document your findings in Table 15.7.

Table 15.7 | Results: Pathological Reflexes

	Right		Left	
	Normal	Abnormal	Normal	Abnormal
Babinski Reflex Response				
Hoffman Reflex Response				

Which spinal pathway do these tests evaluate? ______________________________

____

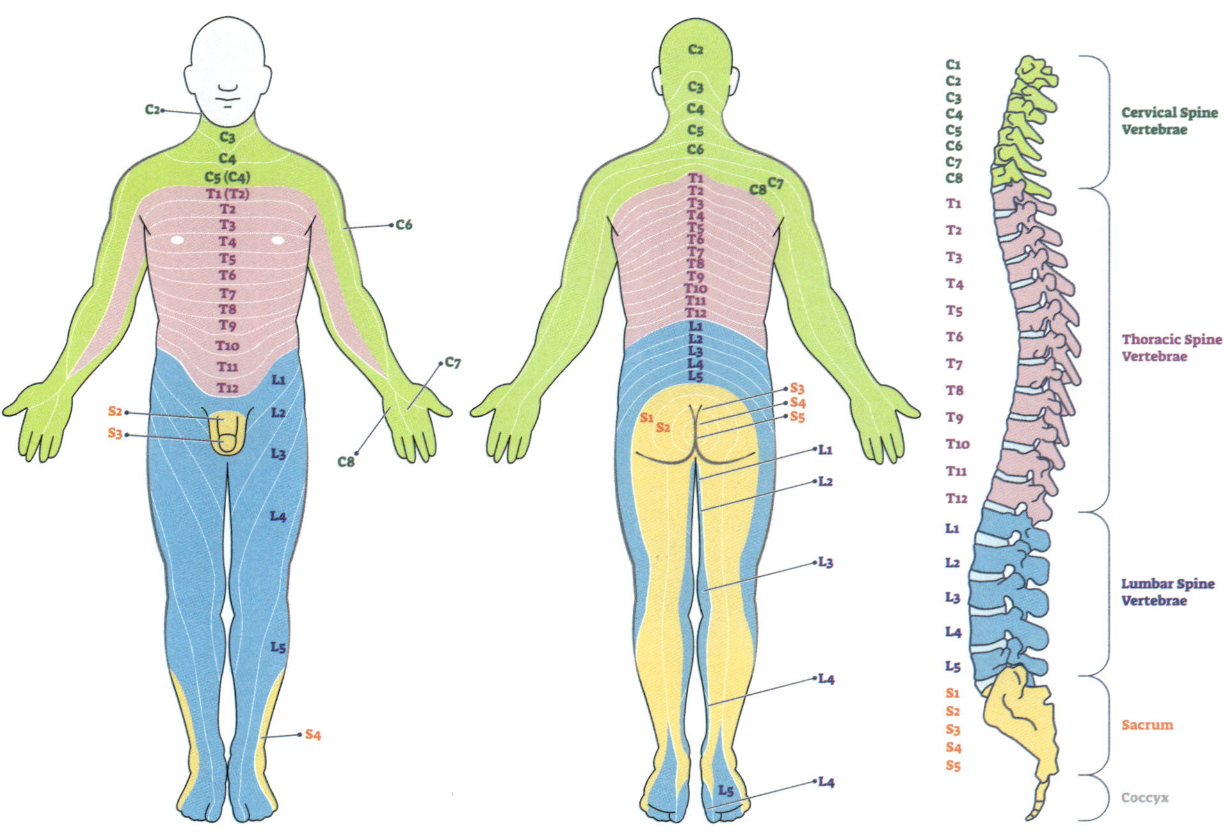

Figure 15.1 | Dermatomes

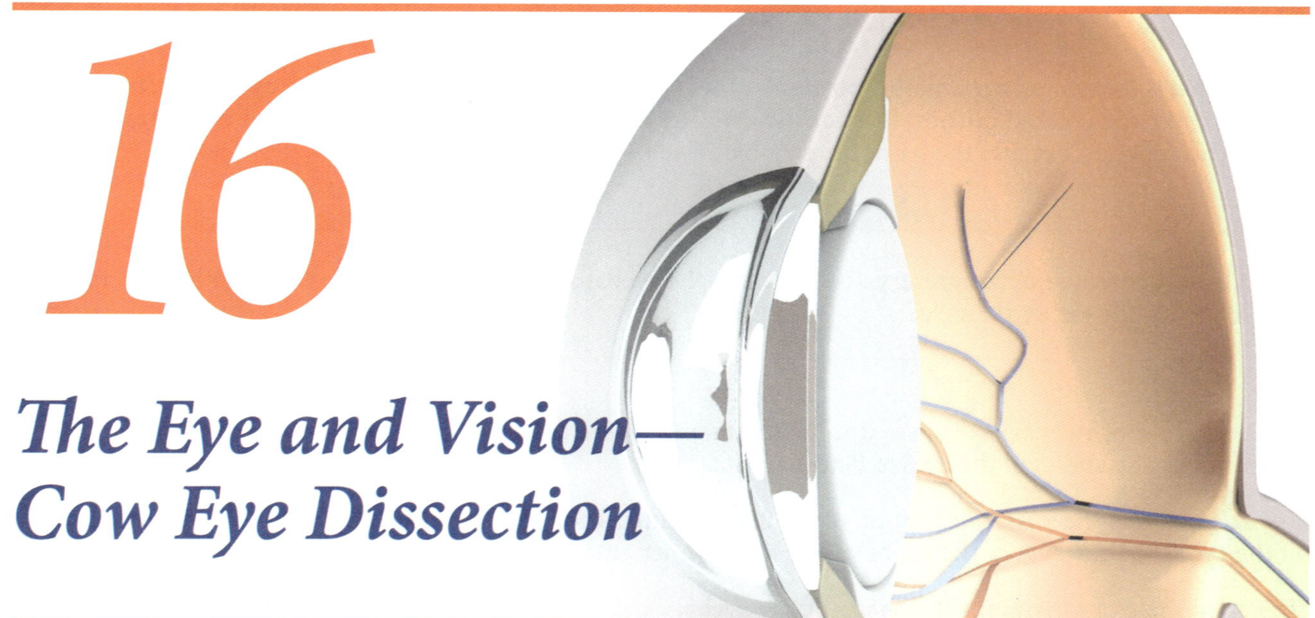

16

The Eye and Vision—Cow Eye Dissection

Materials Needed to Complete this Lab

- Preserved cow eyeball
- Dissecting tray
- Dissecting tools: scalpel, forceps, scissors, probe
- Disposable gloves

Lab Activity

Obtain a preserved cow eyeball and locate as many of the following structures as possible. Use your text or provided materials for reference.

Lab Note: Please use only **one** eyeball per lab table. When finished with the eyeballs, they will be returned to storage in the pail. Wash and dry all dissection trays and tools.

When in the body, the eye is protected by the bones of the orbit as well as layers of fatty tissue. Six ***extrinsic eye muscles*** (four ***rectus*** and two ***oblique***) attach the eye to the bony orbit and make it possible to fix your gaze on a stationary or moving object as well as to coordinate the two eyes so they both look at the same object at the same time. These muscles may or may not be seen on your eyeball specimen. They often will be embedded in the fatty tissue which surrounds it.

A | Conjunctiva

The conjunctivae are clear, moist membranes, which cover the white of the eye and line the upper and lower lids. They are continuous with the outer covering of the eyeball and extend partially into the orbit. For this reason, the conjunctiva will not be found on your specimen since it has been removed from the bony socket. ***Conjunctivitis*** occurs when the conjunctiva becomes inflamed; it is more commonly known as ***pinkeye.***

B | Fibrous Tunic: Sclera and Cornea

The ***cornea*** is located on the anterior portion of the fibrous tunic (the outer layer of the eye). It is a transparent window over the colored portion (iris) of the eye. The cornea can be easily damaged and easily transplanted with little rejection because the cornea has no blood supply.

Lab Note: Because of the preservative in which the eyeballs are stored, the cornea on your lab specimen will be cloudy rather than clear.

The curvature of the cornea causes it to be the main light-refracting structure of the eye. Refraction of light accounts for the ability of the eyes to detect an image.

The ***sclera*** forms the posterior portion of the fibrous tunic. The sclera is more commonly known as the "white" of the eye. The thickness of the sclera varies; it is thinner anteriorly and thicker posteriorly.

The sclera helps to maintain the shape of the eyeball, protect the inner contents, and provide a location for attachment of the extrinsic eye muscles. It contains many sensory nerve endings and blood vessels.

Try to remove the fat from around the exterior portion of the eyeball and notice the sclera beneath it. After you remove some fat, examine the posterior aspect of the eye externally; you should be able to see the round ***optic nerve*** as it leaves the eye.

C | Vascular Tunic (Uvea): Choroid, Iris, and Ciliary Body

The vascular tunic lies beneath the fibrous tunic and includes the ***iris,*** which is the colored part of the eye. The color comes from the pigment ***melanin.*** There are several genes that determine the type and amount of melanin in an individual's eyes. It is closely correlated with skin and hair color. Very little pigment imparts a gray color, next is blue, then brown, then black. Any other color is a variation from additional pigments.

The iris contains autonomic nervous system controlled *constrictor* and *dilator* muscles that control the diameter of the pupil. The ***constrictor muscles*** form a ring around the pupil, and the ***dilator muscles*** radiate away from the edge of the pupil. Constriction of the pupil is parasympathetic, while dilation is sympathetic. The ***pupil*** is simply a hole in the middle of the iris, which allows light into the interior of the eyeball.

Using a scalpel, cut the eye along a coronal (frontal) plane. This will allow the eye to be separated into anterior and posterior sections for easier examination.

Caution—Lab Safety Note: Please use caution when cutting with the scalpel. The preserved specimens are tough and may present difficulty in cutting. If you are having trouble, please consult your lab instructor.

The ***choroid*** is a dark brown, vascular layer lining the sclera. It functions to absorb light rays and prevent them from scattering on the retina. The vessels in this area nourish the retina. Inside your cow eye, you will be able to see the choroid as a thin, black layer lining the sclera. Gently pull it back to separate the two layers from each other.

The choroids of some mammals are lined with a pearly, blue-green pigment called the tapetum lucidum. The ***tapetum lucidum*** collects light at low levels and helps reflect light back onto the retina to improve night vision in some animals. Humans do not have this pigment.

Near the anterior portion of the eye, at the junction between the cornea and sclera, the choroid becomes the ***ciliary body.*** This structure is composed of the ***ciliary muscle*** with folds called ***ciliary processes. Suspensory ligaments*** from the lens attach to the ciliary processes and help center the lens behind the pupil. The ciliary body extends posteriorly into the anterior edge of the neural tunic in an area called the ***ora serrata*** ("saw-toothed mouth"). It is the junction between the neural tunic and ciliary body.

Lab Note: Locate the ora serrata on the eye and predict where the ciliary body might be.

D | The Lens

Located just behind the iris, the **lens** is a transparent, pliable structure made of concentric layers of cells in a dense fibrous capsule. The exterior of the lens is composed of ***elastic fibers;*** the interior contains long, slender cells called ***lens fibers*** filled with ***crystalline proteins.***

The lens is involved in the visual process known as accommodation (the ability to adjust the focal length of the lens for viewing objects near and far away). This involves the action of the ciliary body muscle and suspensory ligaments which alter the shape of the normally elastic lens for accommodation.

Lab Note: When you examine the eye in lab, the lens will appear opaque and hard due to the preservation process.

Loss of transparency of the lens is a *cataract.* Lens elasticity decreases with age and it cannot change shape as readily; thus, as you age, your prescription for eyeglasses or contacts may change.

E | The Neural Tunic: The Retina and Photoreceptors

The ***retina,*** also known as the ***neural tunic,*** is the innermost layer of the eyeball; it is only found in the posterior portion of the eye where it lines the inside of the choroid. It is composed of two layers: a ***pigmented outer layer*** and an ***inner layer*** that contains the eye's photoreceptors, supporting cells, and associated neurons and blood vessels.

The main function of the retina is image formation. It contains sensory ***photoreceptor*** cells known as ***rods*** and ***cones,*** both of which are light sensitive.

The ***rods*** allow vision in dim light, while the ***cones*** are responsible for color vision. The rods form a band along the periphery of the retina.

The cones form a layer on the posterior retinal surface. The highest concentration of cones is found in the ***fovea centralis (central fovea)*** which is a shallow depression at the center point of an oval region called the ***macula lutea.*** There are no rods here; the fovea is where the visual image is directed for sharpest, clearest vision.

The ***visual axis*** is an imaginary line drawn from the center of an object being viewed to the fovea. The visual axis is used to determine how glasses are ground and fitted.

Lab Note: In the lab, the retina will appear as a very thin, delicate, yellowish membrane on the inside of the choroid. Although it would normally be in contact with the entire inside of the posterior portion of the eyeball, manipulation of the eyeball through dissection can cause the retina to tear away and collect in one spot or fall out of the eye altogether.

The retina is dependent upon the choroid for its nourishing blood supply. If the retina should separate from the choroid for a lengthy period of time, the photoreceptor cells would die, and vision could not occur. This condition is known as a *detached retina.*

There are no photoreceptors (rods and cones) in the area just lateral to the central fovea where the ***optic nerve,*** retinal artery, and retinal vein pass through the back of the eye. This area is known as the ***optic disc;*** since there are no photoreceptors for sight there, it creates a ***blind spot*** on the retina.

F | The Cavities of the Interior Eye

The lens separates the eye into two cavities: 1) the ***anterior cavity,*** found between the cornea and ciliary body; and 2) the ***posterior*** or ***vitreous cavity*** between the lens and the retina.

The ***anterior cavity*** is further divided into two smaller chambers—the ***anterior chamber*** (between the cornea and iris) and the ***posterior chamber*** (between the iris and lens). The anterior cavity (and both chambers) is filled with a fluid called ***aqueous humor.*** It is a clear, watery liquid that is continually replaced and filtered in order to nourish the lens and cornea and to remove their waste products. The presence of the aqueous humor provides cushioning and also helps to stabilize the position of the retina to ensure that the photoreceptors are pushed tightly against the pigmented layer of the retina.

The ***posterior (vitreous) cavity*** is much larger than the anterior cavity. It is filled with a clear, gelatinous material called ***vitreous humor (body).*** This material helps maintain the position of the retina and keeps the eyeball tissues from collapsing. Unlike the aqueous humor which is produced continually, vitreous humor is formed during fetal development and maintained throughout life.

Lab Note: Although you will be able to examine the vitreous humor, the aqueous humor leaks out of the eyeball when it is cut open. Movement of the vitreous humor in a cut eyeball often pulls the retina away from its contact with the posterior portion of the eyeball.

Required Structures of the Eye

After completion of the lab, you should be able to identify and correctly spell the following eyeball structures on the preserved/dissected eyes and eye model.

- Cornea
- Sclera
- Iris
- Pupil
- Choroid
- Ciliary body
- Ora serrata
- Tapetum lucidum
- Lens
- Retina
- Macula lutea
- Central fovea (if available)
- Optic disc
- Optic nerve
- Anterior cavity
- Anterior chamber
- Posterior chamber
- Posterior cavity
- Aqueous humor
- Vitreous humor
- Extrinsic eye muscles

Laboratory 16 Review Questions
Eyeball Structures

Name: ______________________________

1 | For each of the following structures, briefly describe their function.

a | Choroid:

b | Sclera:

c | Lens:

d | Ciliary body:

2 | Name the two types of photoreceptors and indicate where they are located in the eye.

3 | What creates the blind spot at the back of the retina?

4 | List the structures found in the following layers of the eye.

a | Fibrous tunic:

b | Vascular tunic:

c | Neural tunic:

Match the following:

5 |	__________ Area of high cone concentration	*a* |	Cornea
6 |	__________ Avascular light-refracting structure	*b* |	Conjunctiva
7 |	__________ Iris pigment	*c* |	Vitreous humor
8 |	__________ Attaches lens to ciliary body	*d* |	Macula lutea
9 |	__________ Found in posterior cavity	*e* |	Suspensory ligament
10 |	__________ Clear membranous lining of eyelids	*f* |	Melanin

17

The Eye and Vision—Vision Testing

Materials Needed to Complete This Lab

- Snellen Eye Chart
- Astigmatism Chart
- Meter stick
- Ichikawa Color Blind Test Books
- Afterimage test cards
- Peripheral vision disks
- Lab book

You will be performing a variety of vision tests in lab. Your results will be recorded in the tables at the end of this lab exercise.

Lab Note: You will need to work with a partner so that you may test each other's vision and record the results.

For an image to be correctly focused on the retina, several processes must work together:

1 | ***Constriction*** of the pupil

2 | ***Accommodation*** of the lens

3 | ***Convergence*** of the eyes so that they are both focused on the same object

4 | ***Refraction*** of light through the cornea

5 | ***Activation*** of photoreceptors

Keep in mind that the image must be properly focused on the retina, and it must be detected by the ***photoreceptors*** (rods and cones). Then a nerve impulse is sent via the ***optic nerves,*** through the ***optic chiasma,*** the ***optic tracts,*** the thalamus, and to the ***primary visual cortex*** where the image is perceived. Finally, the message is sent to the nearby ***visual association areas*** of the cerebral cortex where the image is interpreted.

A | Visual Acuity

The visual acuity represents the sharpness or clarity of vision. For the clearest vision, the light rays must be focused on the ***central fovea*** where the highest concentration of cones is located.

The ***Snellen Eye Chart*** is a common test used to test for visual acuity.

Lab Activity

1 | Locate the Snellen Eye Chart on the wall or door of the lab room. Stand at a distance of 20 feet from the chart. (**Note:** There should be marks on the floor or wall indicating 20 feet. If you cannot find them, please inform your lab instructor.)

Lab Note: The table where you will record your results has a box for corrected or uncorrected vision testing. If you **do not** wear glasses or contacts, you will record your results in the "uncorrected" box. If you **do** wear glasses or contacts, you may leave them on for your vision testing; in that case, place your results in the "corrected" box. If you choose to remove your corrective lenses, then mark your results in the "uncorrected" box.

2 | Cover one eye and read the letters on the chart from top to bottom until you can no longer distinguish them correctly. Have your lab partner record your results in Table 17.1 in ***your*** lab book.

Lab Note: Record the numbers on the left side of the eye chart. This gives the visual acuity index when read from 20 feet away.

3 | The last line of letters that can be read with up to two mistakes is the limit of the visual acuity.

4 | Repeat with the other eye, recording the results in Table 17.1.

A score of 20/20 vision means that you are able to see at 20 feet what a "normal" eye can see at 20 feet. The top number represents your visual distance while the bottom number represents the visual distance of an individual with "normal" vision.

20/100 vision means that you can see at 20 feet what a normal eye can see at 100 feet away. (Remember the top number is ***your*** distance and the bottom number represents a person with normal vision). Not so good!

20/15 vision means that you can see at 20 feet what a normal eye could only see clearly at 15 feet. This is considered better than normal vision. However, this can be misleading because the person who has better far vision may have trouble focusing on near objects.

Lab Tip: When trying to interpret your score, keep in mind that the top number represents ***your*** vision, while the bottom number represents "normal" eyes.

B | Astigmatism

In order for an image to be properly focused onto the retina, it must be properly ***refracted*** (bent) by the refractive structures of the eye (cornea and lens).

Astigmatism occurs where there is an abnormal curvature of either the cornea or lens, resulting in improper refraction and image formation on the retina.

The astigmatism test chart resembles a wheel with a series of numbered "spokes" extending from the center. The lines that radiate out from the center seem darker and bolder to the parts of the eye which have an astigmatism. Light rays will have the greatest refraction (bending) where the cornea has the most curvature and will, therefore, reach their focal point in front of the retina and seem blurred. The eye will try to compensate by trying to focus the light from the flattened area of the cornea (the part with the astigmatism). The normally curved portion of the cornea will then over-refract the light.

Lab Activity

Lab Note: The table where you will record your results has a box for corrected or uncorrected vision testing. If you **do not** wear glasses or contacts, you will record your results in the "uncorrected" box. If you **do** wear glasses or contacts, you may leave them on for your vision testing; in that case, place your results in the "corrected" box. If you choose to remove your corrective lenses, then mark your results in the "uncorrected" box.

1 | Locate the Astigmatism Chart on the wall or door of the lab room. This chart resembles a wheel with a series of paired spokes extending from the center.

Stand at a distance of 10 feet from the chart. (**Note:** There should be marks on the floor or wall indicating 10 feet. If you cannot find them, please inform your lab instructor.)

2 | Cover one eye and stare at the center of the chart.

3 | The lines should all appear of equal darkness and be straight. If not, an astigmatism may be present. Record your results in Table 17.1.

4 | Repeat with the other eye and record your results in Table 17.1.

C | Accommodation

In order to focus an image on the retina, we must change the shape of the lens; this is known as ***accommodation.*** Accommodation for near vision causes the lens to become rounder while accommodation for far vision causes it to flatten out.

When focusing on a nearby object, the ciliary body contracts, the suspensory ligaments relax, and the lens becomes more rounded. The opposite occurs for far vision.

As the light refracts through the lens, it pinpoints the image to an area on the retina. The pinpointed area is the ***focal point.*** The distance between the center of the lens and the focal point is called the ***focal distance.*** The focal distance depends upon two factors: 1) the distance between the lens and the object seen; and 2) the lens shape. The fatter the lens, the more refraction there is (a fatter lens refracts more; therefore, the focal distance is shorter).

Nearsightedness (also known as ***myopia***) occurs if the eyeball is too long, and the visual image focuses in front of the retina. Someone who is myopic can see well up close but is unable to flatten the lens enough to see well far away.

Farsightedness (also called ***hypermetropia*** or ***hyperopia***) is a short eyeball, which causes an image to focus behind the retina. Objects viewed at a distance are clear but the lens is unable to accommodate for near vision.

As the lens loses elasticity over time, the near vision is affected first. This results in ***presbyopia***—a type of farsightedness that occurs with aging of the lens.

This test measures the inner limit of clear vision.

Lab Activity

1 | Obtain a meter stick and place it on the lab table, with one end near the edge of the table. Leave enough room so that you may rest your chin on the edge of the table top in front of the meter stick.

2 | Have your lab partner hold your lab book at the one-meter mark on the meter stick. Close one eye and focus on the word "HOME" in the box below. (The lab book may need to be bent slightly in order to clearly see the word in the box.)

3 | Have your partner slowly move the book closer toward your eye until the word "HOME" is out of focus; then move the book very slowly backward until it is focused again. This is your near point of vision. Record the distance in your results in Table 17.1.

4 | Repeat with the other eye.

HOME

D | Blind Spot

At the back of the retina, there is an area where the optic nerve, retinal artery, and retinal vein pass through the eye. This area is known as the ***optic disc*** or ***blind spot*** because it has no photoreceptors. We cannot see images that strike the blind spot.

Lab Activity

1 | Close your right eye. Hold your lab manual with your left hand at arm's length from your face.

2 | Stare at the circle (•) in the following box.

3 | ***Slowly*** bring the manual closer to your face while you remain staring at the circle. At some point the cross (+) in the box will disappear as its image falls on the blind spot.

4 | Now close your left eye and hold the lab manual in your right hand. Repeat Steps 1–4, but stare at the cross (+) and observe if the circle (•) disappears as its image falls on the blind spot.

5 | Record your results in Table 17.1.

E | Negative Afterimages

If the red, green, and blue cones are all stimulated simultaneously, white light is seen. Staring at objects for short or long periods of time can cause the image to seem to appear on the retina after the line of sight is moved. If the image retained on the retina matches the image that was seen, then a **positive afterimage** has been produced. Staring at objects for short periods of time or looking at a bright light (e.g., flashbulb) will produce a positive afterimage on the retina. A **negative afterimage** is produced when an object is stared at for an extended period of time. The unmoving image that is being viewed is thought to tire the photoreceptors. As a result, there is an optical illusion that causes a reversal of the colors in the image.

Lab Activity

1 | Use the brightly colored red and green crosses (+) provided.

2 | Without blinking, stare at the colored (+) for 30 seconds.

3 | After 30 seconds, immediately close your eyes and concentrate on the image that is formed on your eyelids.

4 | Note whether or not the colors formed while your eyes were closed were reversed from the actual colors on the (+).

5 | Record your results in Table 17.1.

F | Field of Vision

The ability to see peripherally varies between individuals and may be affected by disease or damage to the eyes. A normal field of vision for one eye is 130 degrees; for both eyes, a normal field of vision is 180 degrees.

Lab Activity

This activity requires three individuals to perform the measurement—the individual being tested (**subject**), a **tester** who makes measurements, and a **recorder** who is responsible for recording the results reported by the tester.

Instructions for the Individual (Subject) Being Measured

1 | The individual (subject) who is being tested sits at a lab table, folds down the handles of the vision disk, and places it against his/her forehead. In the correct position, it should resemble a giant visor.

2 | Next, the subject folds down the focus marker so that it hangs straight down beneath the disk.

3 | The subject should grasp the handle with both hands, spread his/her elbows, and rest them on a tabletop for stability. The subject's hands should be approximately level with his/her mouth.

4 | The subject should look straight ahead under the disk and focus on the hole in the focus marker. This focusing helps prevent undesired eye movement.

Instructions for the Tester Who Is Making the Measurements

1 | The tester should stand behind the subject. With the knob on the arm of the Field of Vision Disk, the tester moves the arm to the right side of the subject's head.

2 | The tester places one sight card into the slit on the arm, making sure that the subject does not see the card before the test.

Measuring the Field of Vision

1 | Starting with the vision disk on the right side of the subject's head, the tester slowly swings the arm with a sight card in place towards the front of the subject.

2 | The recorder watches to see that the subject's eyes do not move.

3 | The subject announces when the card is first seen.

4 | The tester determines the reading at the apex of the arrow cut in the arm and tells the recorder the number.

5 | The recorder records this number in Table 17.1 as the limit of the right side Field of Vision.

Measuring the Reading Field of Vision

1 | The tester continues to slowly move the arm towards the front of the subject, and the recorder continues to watch the eyes of the subject to ensure they remain focused on the marker.

2 | The subject says the letter aloud as soon as it can be read.

3 | The tester reads the measurement and the recorder records the measurement in Table 17.1 as the limit of the right side Reading Field of Vision.

Repeat both of the above tests by starting with a new sight card and the arm on the left side. Place those results in Table 17.1 as the left-side measurements.

The sum of the right and left side measurements for the Field of Vision and the Reading Field of Vision represents the respective ranges in degrees.

Visual Function Test Results

Name: ____

Table 17.1 | Visual Test Results

Lab Test	Right Eye	Left Eye
Snellen (visual acuity) test	**Corrected:** **Uncorrected:**	**Corrected:** **Uncorrected:**
Astigmatism	**Corrected** ☐ **Normal** ☐ **Not Normal** **Uncorrected** ☐ **Normal** ☐ **Not Normal**	**Corrected** ☐ **Normal** ☐ **Not Normal** **Uncorrected** ☐ **Normal** ☐ **Not Normal**
Accommodation (near point)	**Corrected:** ____________ **cm** **Uncorrected:** ____________ **cm**	**Corrected:** ____________ **cm** **Uncorrected:** ____________ **cm**
Blind spot	☐ **Found** ☐ **Not Found**	☐ **Found** ☐ **Not Found**
Negative afterimage	☐ **Found** ☐ **Not Found**	☐ **Found** ☐ **Not Found**
Peripheral Vision Disk: **Field of Vision Measurement** **Reading Field of Vision**	________ **degrees** ________ **degrees**	________ **degrees** ________ **degrees**

Remarks on Results:

G | Color Vision and Deficiency

On the retina the photoreceptors responsible for detecting color are three types of ***cones.*** Cones are specialized for high intensity light vision and allow us to distinguish between colors. Color vision is known as ***photopic vision.*** The different types of cones are sensitive to wavelengths of light rays that correspond to the colors red, blue, and green. Considerable overlap exists and stimulating the different types of cones in different numbers will allow the brain to detect various shades of colors. If all three types of cones are stimulated equally, white light is perceived.

On the other hand, the retinal photoreceptors known as ***rods*** are sensitive to low light levels allowing sight in dim light conditions. They are unable to perceive color but discriminate between shades of darkness and allow us to perceive shapes and movements. This night vision, or "black-and-white" vision is known as ***scotopic vision.***

A person who is ***color deficient*** is unable to perceive one or more of the three primary colors (red, blue, or green). The degree of the deficiency varies from confusion regarding the colors to a total inability to perceive color.

Lab Note: The term "color blindness" is no longer commonly used because of the negative connotation associated with the word "blindness."

The most common type of color deficiency is due to a defect in the red cones and is known as red-green deficiency. The wavelengths for red and green light stimulate the same concentrations of cones; therefore, there is confusion as to which color is which.

Red-green color deficiency is a sex-linked genetic trait (passed from mother to son) and affects approximately 9% of the male population and less than 1% of the female population. A sex-linked trait is passed along the X (female) chromosome. Each baby gets an X chromosome from the mother and an X or Y (male) chromosome from the father. An XX combination produces a female; an XY a male.

Since the color deficient trait is passed on the X chromosome, a male needs only one gene for color deficiency (which he will receive from his mother's X chromosome), but a female would need two genes for color deficiency (one received from her mother and one from her father). If a female has one gene for color deficiency and one normal gene, she will not be color deficient (but will be a carrier of the color deficient gene). Thus, red-green deficiency is rare in females.

This type of color deficiency is known as ***congenital*** because it is inherited and present at birth. With this type of deficiency, the red or green cones are missing or nonfunctional. It affects both eyes.

However, color deficiency can also occur because of damage to the nerve or brain. This type of deficiency is known as ***acquired*** and may not necessarily affect both eyes, dependent upon the source of the damage.

An individual with nonfunctional or missing red cones has ***protanopia.*** Defective or missing green cones causes ***deuteranopia.***

Lab Activity (Main Campus Labs)

Caution: These test books are extremely expensive. **Do not** touch the color plates with your fingertips as the oils from the hand will distort the ink and destroy the plates.

1 | Obtain the Ichikawa color charts. There are two books in the set—one which tests for congenital color deficiency and one for acquired deficiency.

2 | Have your partner hold the book approximately 12 inches from you as you read the numbers on the plates. There are different directions and scoring methods for the two types of books:

Congenital Color Deficiency Test

1 | Obtain the book labeled for congenital color deficiency testing. Plates 1–4 are demonstration plates only. You may skip these plates.

2 | ***Keeping both eyes open,*** begin the test on plate five. Have your partner record your results in Table 17.2 in your lab book as you proceed through the plates.

3 | Plates 5–14 are used to initially screen for the possibility of a congenital color deficiency. In Table 17.2, fill out the results (use the "Screening Series" table) for Plates 5–14 by **circling the number read by the person being tested. When two numbers are identified on the test plate, circle only the number which is seen *best.***

4 | After completing Plate 14, stop and count the number of circles in each column (Normal and R-G Defect). If there are 8 or more numbers circled in the "Normal" column, then the testing is done, and that individual has normal (non-color deficient) vision.

Continue to Step 5 only if the screening test indicates a possible red-green defect.

5 | If there are less than 8 numbers circled in the "Normal" column and more numbers circled in the "R-G Defect" column, then the individual may have a red-green defect. In that case, continue the testing with Plates 15–19 and record those results in Table 17.3 as "Classification Series." Count the number of circles in each column to determine if the individual has defective red (protanopic) or green (deuteranopic) cones.

Acquired Color Deficiency Test (Main Campus Labs)

Note that this is a different scoring method than for congenital deficiency.

1 | Obtain the book for acquired color deficiency. Plates 1 and 2 are test plates. You may skip those.

2 | For acquired color deficiency, ***test each eye individually.*** (There may be damage to the photoreceptors in one eye and not the other.) Your results will be recorded in Tables 17.4 (right eye) and 17.5 (left eye).

3 | Beginning with Plate 3 and continuing through Plate 12, **circle *each* number that is read correctly and place an "X" on each number that cannot be read or is read incorrectly.**

4 | After completion of Plate 12, count the number of X's in each column. Many X's on the numbers marked BY means blue-yellow defect; many X's on the RG means red-green defect. X's on the numbers marked S means scotopic vision.

Score Sheets and Results for Congenital Color Deficiency (Main Campus Labs)

Instructions: Use the table below to initially screen for potential color deficiency. After completing the tests, count how many numbers have been circled in each column. *If the "normal" column has 8 or more numbers circled, then this test is done and vision is normal.*

Place a circle around the number seen in each plate. If two numbers are observed, circle ***only*** the one which is better read.

Screening Series of Plates

Name: ____

Table 17.2 | Plate Numbers 5–14

Plate Number	Normal	R-G Defect
5	3	8
6	2	9
7	4	Invisible
8	7	4
9	8	7
10	4	3
11	2	4
12	7	5
13	8	Invisible
14	3	6
Total Count		

Instructions: Use the table below ***only*** if the results in the Screening Series indicate (if there are less than 8 numbers in the "Normal" column or if there are many numbers circled in the "R-G Defect" column).

Place a circle around ***each*** number observed. Then count how many circles are in each column to determine red or green deficiency.

Interpretation of results for congenital color deficiency screening (mark one box):

☐ Normal

☐ Protanopia

☐ Deuteranopia

Classification Series of Plates

Table 17.3 | Plate Numbers 15–18

Plate Number	Protan	Deutan
15	8	3
16	5	7
17	4	8
18	9	4
Total Count		

Score Sheets and Results for Acquired Color Deficiency (Main Campus Labs)

Instructions: Use the following tables to screen for acquired color deficiencies. Test each eye individually. For each plate, circle the number read correctly and place an X on any number that cannot be read or is read incorrectly. X's are diagnostic: X on BY numbers indicates a blue-yellow defect; X on RG numbers indicates red-green defect; X on S numbers indicates scotopic vision.

For normal results, ***all*** numbers of all plates should be read (circled).

Name: ____

Table 17.4 | Right Eye

Plate Number	Right Eye	
3	2 *BY*	4 *BY*
4	6 *BY*	7 *BY*
5	3 *BY*	2
6	6 *BY*	3
7	5 *BY*	9 *S*
8	9 *BY*	8 *RG*
9	5 *BY*	2 *RG*
10	2 *BY*	6 *RG*
11	3 *BY*	5 *RG*
12	4 *RG*	3 *S*

Table 17.5 | Left Eye

Plate Number	Left Eye	
3	2 *BY*	4 *BY*
4	6 *BY*	7 *BY*
5	3 *BY*	2
6	6 *BY*	3
7	5 *BY*	9 *S*
8	9 *BY*	8 *RG*
9	5 *BY*	2 *RG*
10	2 *BY*	6 *RG*
11	3 *BY*	5 *RG*
12	4 *RG*	3 *S*

Interpretation of results for acquired color deficiency screening (mark one box):

☐ Normal

☐ Blue-yellow

☐ Red-green

☐ Scotopic

Congenital Color Deficiency Test (Akron Campus Labs)

Instructions: Beginning with Plate 1, place a circle around the number seen in each plate. If no number is seen, then no mark needs to be made in the column.

Table 17.6 | Screening Series of Plates. Name: ___________________________

Plate	Normal Person	R-G Defect	Total Color Blindness
1	12	12	12
2	8	3	Not seen
3	29	70	Not seen
4	5	2	Not seen
5	3	5	Not seen
6	15	17	Not seen
7	74	21	Not seen
8	6	Not seen	Not seen
9	45	Not seen	Not seen
10	5	Not seen	Not seen
11	7	Not seen	Not seen
12	16	Not seen	Not seen
13	73	Not seen	Not seen
14	Not seen	5	Not seen
15	Not seen	45	Not seen
Total count (of numbers circled)			

Use plates 16 and 17 only if you have 9 or less numbers circled in the Normal column.		Protan		Deutan		
		Strong	Mild	Strong	Mild	
16	26	6	(2) 6	2	2 (6)	
17	42	2	(4) 2	4	4 (2)	
Total Count						

If there are 13 or more numbers circled in the Normal column, then there is no color deficiency and the test is complete. If there are 9 or less numbers circled in the Normal column, then there may be a deficiency. Complete Plates 16 and 17 to determine protanopia or deuteranopia.

Results for congenital color deficiency screening (mark one box):

☐ Normal

☐ Protanopia

☐ Deuteranopia

Laboratory 17 Review Questions

The Eye and Vision—Vision Testing

Name: ______________________________

Instructions: For each of the following statements, record your answer in the spaces on the bottom of the page.

1 | The sharpness of vision is known as ___(1a)___ and can be evaluated by using the ___(1b)___ eye test.

2 | A person who has 20/50 vision has ___(better than) (worse than)___ normal vision.

3–4 | An astigmatism results from the abnormal curvature of either the __________ or the __________.

5 | Photoreceptors responsible for dim light vision are __________.

6 | Photoreceptors responsible for color vision are __________.

7 | A person who lacks functional green cones has __________ which is a type of red-green color deficiency.

8 | A person who lacks functional red cones has __________ which is also a type of red-green color deficiency.

9–10 | Color deficiencies that are inherited are known as ___(9)___, whereas those due to eye or nerve damage are ___(10)___.

11 | The ability of the lens to change its shape to aid in near or far vision is called __________.

12 | The __________ is an area of the retina in which there are no photoreceptors.

13 | Another word for color vision is __________ vision.

14 | Black-and-white vision is also known as __________ vision.

15 | An individual who can see objects nearby but cannot see objects at distance has a condition known as __________ or nearsightedness.

______________________________ 1a.

______________________________ 1b.

______________________________ 2.

______________________________ 3.

______________________________ 4.

______________________________ 5.

______________________________ 6.

______________________________ 7.

______________________________ 8.

______________________________ 9.

______________________________ 10.

______________________________ 11.

______________________________ 12.

______________________________ 13.

______________________________ 14.

______________________________ 15.

18

Structures of the Ear

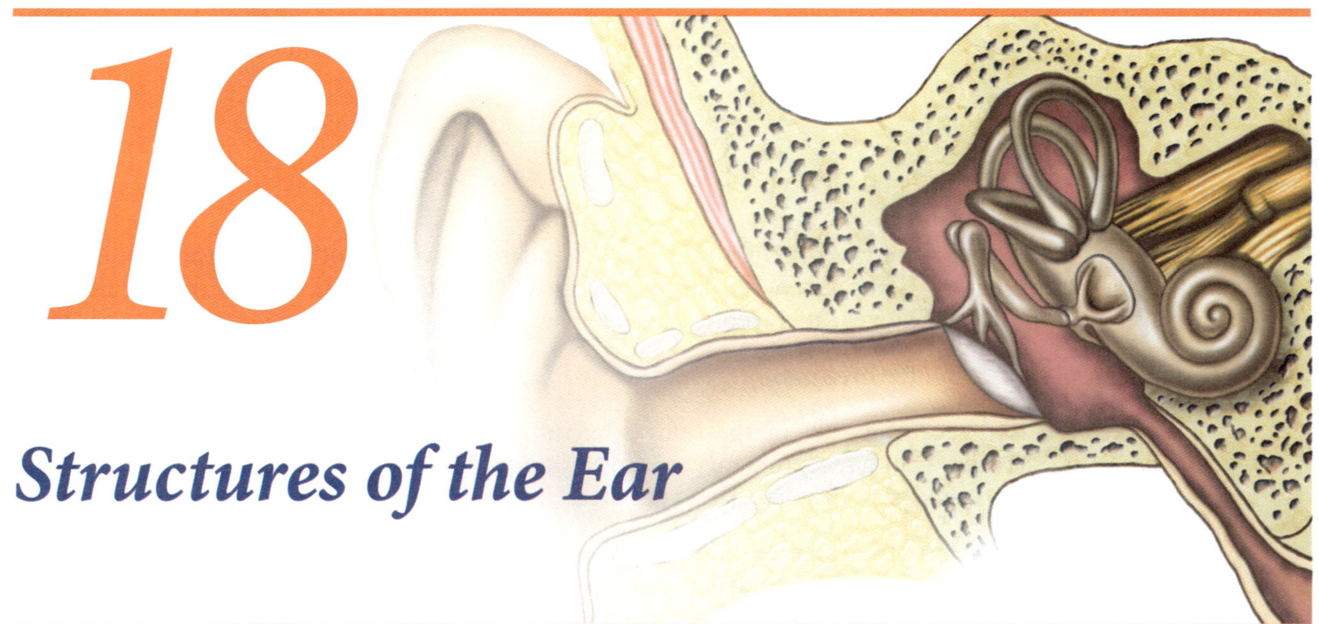

The structures of the ear are divided into three main regions: 1) the ***external ear;*** 2) the ***middle ear;*** and 3) the ***inner ear.***

External (Outer) Ear

The external ear consists of the auricle (pinna), the external auditory canal, and the eardrum. The **auricle** is the cartilaginous structure that extends from lateral surfaces of your head—what we commonly think of as our "ears." It collects sound waves and directs them inward.

The sound waves travel through the **external auditory canal (meatus)** which is a tubular structure about 1" long between the auricle and eardrum. It contains **ceruminous glands** that secrete **cerumen** (earwax); cerumen helps keep dust and foreign objects out of the ear.

The **tympanic membrane** (eardrum) rests at the end of the external auditory canal. It is a thin, semitransparent membrane connecting the external auditory canal and the middle ear. When sound waves reach the tympanic membrane, it begins to vibrate and transmit the waves to the auditory ossicles of the middle ear.

Middle Ear

Beyond the distal end of the tympanic membrane lies the middle ear. It is an air-filled cavity carved from the temporal bone. The middle ear is separated from the external ear by the tympanic membrane and from the inner ear by a bony partition which contains two additional thin membranes—the **oval window** and the **round window.**

Crossing the middle ear are three extremely small bones—the **auditory ossicles.** The first ossicle, the **malleus** ("hammer"), is attached to the tympanic membrane; the **incus** ("anvil") is the middle bone and is attached to the malleus and the **stapes.** The stapes ("stirrup") is the third auditory ossicle, and it is attached at its distal end to the oval window. Sound waves from the tympanic membrane are transmitted through the middle ear by the ossicles and eventually to the oval window.

Also contained in the middle ear is the **pharyngotympanic tube** (also known as the **auditory tube** or **eustachian tube**). This tube opens to the middle ear and runs diagonally to the nasopharynx (upper portion of the throat). It helps maintain proper air pressure in the middle ear so that the tympanic membrane can vibrate freely.

Two tiny skeletal muscles are attached to the ossicles. They are the **tensor tympani muscle** and the **stapedius muscle.** When the ear is exposed to extremely loud noises, these muscles contract and help prevent damage to the hearing receptors.

Internal (Inner) Ear

The inner ear contains a series of bony and membranous canals and is involved in both hearing and equilibrium. It is called the **labyrinth** ("maze") because of the complex shape of these canals. The inner ear is located deep in the temporal bone to protect the delicate receptors for hearing.

The **bony labyrinth** is divided into three main regions:

1 | The **vestibule** is the central part of the bony labyrinth. It contains the **utricle** and **saccule,** two membranous sacs containing receptors involved in static equilibrium (keeping track of where the head is in space); they are essential for posture and balance.

2 | Behind the vestibule are three **semicircular canals** extending at right angles to each other. Inside the bony canals are corresponding membranous **semicircular ducts.** At one end of each of the canals are enlarged swellings called the **ampullae.** The ducts and ampullae communicate with other structures of the inner ear and have receptors that respond to rotational movements of the head. They are involved in dynamic equilibrium.

3 | The **cochlea** ("snail") is a bony, spiral-shaped canal extending from the vestibule. Some of the structures contained in the cochlea include a membranous **cochlear duct,** the **spiral organ (organ of Corti), perilymph, endolymph,** and the **tectorial** and **basilar membranes.** The spiral organ contains **hair cells,** the receptors for hearing. Exposure to loud sounds can permanently damage the hair cells, resulting in hearing loss. The entire structure of the cochlea is designed to transmit vibrations and pressure waves that will stimulate the receptors of the spiral organ.

Once the receptors for either hearing or equilibrium are activated, the signals are sent to the brain through the **vestibulocochlear nerve** (8th cranial nerve). It has two branches—the **vestibular nerve** (branch), conducting impulses related to equilibrium and the **cochlear nerve** (branch), conducting impulses for hearing.

Required Structures of the Ear

After completion of the lab, you should be able to identify and correctly spell the following ear structures on the ear model(s).

- External ear
- Middle ear
- Inner ear
- Auricle (pinna)
- External auditory canal (meatus)
- Tympanic membrane
- Auditory ossicles (malleus, incus, stapes)
- Pharyngotympanic tube (auditory tube, eustachian tube)
- Oval window
- Round window
- Vestibule
- Semicircular canals
- Cochlea
- Vestibulocochlear nerve

19

Cranial Nerve Testing

There are twelve pairs of cranial nerves that attach directly to the ventral surface of the brain, classifying them as part of the ***peripheral nervous system,*** or ***PNS.*** The functions of each cranial nerve differ and may include one or a combination of the following:

- **Special sensory:** These nerve fibers carry the sensations of sight, smell, taste, hearing, or balance.
- **Somatic sensory:** These nerve fibers carry general (somatic) sensory information that is detected through the skin such as pain, temperature, touch, pressure, or vibration. Proprioception is also a somatic sense.
- **Visceral sensory:** These nerve fibers carry sensory information from receptors monitoring the internal environment and provide important information for regulating autonomic function.
- **Somatic motor:** These nerve fibers carry commands to skeletal muscle and are under voluntary control.
- **Visceral motor:** Often called **autonomic,** these fibers carry involuntary commands to internal organs and glands.

Cranial nerve testing is an important clinical assessment tool because it can provide clues to determine if one has a problem within the nervous system. Most cranial nerves are associated with the senses, both general and special. In this lab you will assess some or all function(s) of each cranial nerve using standardized tests. Often, one test can be used to assess multiple nerves. You will be working in pairs when performing these tests. Document your findings of each in the Results Tables accompanying each activity.

Materials Needed to Complete This Lab

- Vials of various fragrances
- Penlight
- Cotton balls or Q-tips
- Gloves
- Containers of lemon juice, salt water, sugar water
- Tuning forks
- Rotating stool or chair

Lab Activity 1 | Testing Smell—Olfactory Nerve: Cranial Nerve I

Cranial nerve I emerges directly from the cerebrum and carries special sensory information for the sense of smell. The tissue lining the roof of the nasal cavity and parts of the nasal septum, called olfactory epithelium, is embedded with olfactory receptors. The axons that form from these penetrate the cribriform plate of the ethmoid bone and form the olfactory nerves that enter the olfactory bulbs then olfactory tracts which travel to the olfactory cortex in the temporal lobes.

Materials Needed to Complete This Lab

- Vials of substances with common fragrances

Procedures

1 | **Directions for testing the olfactory nerve:**

a | Ask your partner to close their eyes and plug one nostril.

b | Hold a fragrance vial under one nostril and ask them to identify the smell.

c | Repeat with the other nostril.

d | Repeat these tests using different fragrance vials.

e | Record your results in Table 19.1.

Loss of the sense of smell may indicate damage to the olfactory nerve.

Table 19.1 | Smell Test

Put the name of the scent in the boxes below.	Right Nostril		Left Nostril	
	Odor Detected	Odor Not Detected	Odor Detected	Odor Not Detected
Scent 1:				
Scent 2:				
Scent 3:				
Scent 4:				

Lab Activity 2 | Testing Vision and the Eye—Optic Nerve: Cranial Nerve II; Oculomotor Nerve: Cranial Nerve III; Trochlear Nerve: Cranial Nerve IV; Abducens Nerve: Cranial Nerve VI

Cranial nerve II emerges from the diencephalon and carries special sensory information for vision. Photoreceptors, called rods and cones, are located in the retina and form the optic nerves. The optic nerves travel through the optic canals, converge at the optic chiasm, and continue as optic tracts as they travel to the thalamus then reach the visual cortex in the occipital lobe.

Cranial nerve III arises from the midbrain and carries both somatic and visceral motor fibers to control eye movements. The somatic motor fibers innervate four of the six pairs of extrinsic eye muscles that allow for voluntary movements of the eye, as well as the levator palpebrae superioris muscle that raises the upper eyelid. The visceral motor (autonomic) fibers control the intrinsic eye muscles that regulate the diameter of the pupil.

Cranial nerve IV arises from the midbrain and carries somatic motor fibers to the fifth pair of extrinsic eye muscles that allow us to look down and to the side.

Cranial nerve VI arises from the pons and carries somatic motor fibers to the sixth pair of extrinsic eye muscles that allow us to look to the side.

Materials Needed to Complete This Lab

- Penlight

A | Testing the Optic Nerve (CN II)— A Measure of Visual Acuity or Sharpness of Vision

Refer to Table 17.1 (page 115) for your visual acuity results.

B | Directions for Testing the Somatic Motor Functions of the Oculomotor, Trochlear, and Abducens Nerves (CNs III, IV, VI)

1 | Ask your partner to sit and keep their head still.

2 | Holding the penlight like a stick, move it up and down in an "H" pattern while your partner uses their eyes to follow the penlight.

3 | Then ask your partner to blink their eyes.

4 | Record your results in Table 19.2.

Table 19.2 | Coordination of Eye Movements—Cranial Nerves III, IV, VI

	Right Eye		Left Eye	
	Normal	Abnormal	Normal	Abnormal
Ability to Follow H Pattern				
Blinking				

Inability to coordinate eye movements may indicate damage to one or more of these nerves. Complaints of eye pain, droopy eyelids, or double vision may be signs of damage to oculomotor nerve.

C | Directions for Testing the Visceral Motor Function of the Oculomotor Nerve (CN III)—Assessing the Direct Pupillary Reflex and the Consensual Pupillary Reflex

1 | Ask your partner to sit and keep their head still.

2 | Use the penlight to shine the light into the right eye and observe the pupils.

3 | Pupillary constriction in the right eye indicates a normal direct pupillary reflex; pupillary constriction in the left indicates a normal consensual reflex.

4 | Repeat with the left eye.

5 | Record your results in Table 19.3.

Table 19.3 | Pupillary Reflexes: Cranial Nerve III

	Pupillary Constriction	No Pupillary Constriction
Right Eye—Direct		
Left Eye—Consensual		
Left Eye—Direct		
Right Eye—Consensual		

For a normal direct pupillary reflex, you should see the pupil constricting in the eye you shine the penlight. For a normal consensual pupillary reflex, the opposite pupil should also constrict at the same time. Both pupils should constrict for a normal response. Lack of pupillary constriction may indicate damage to the oculomotor nerve or to the optic nerve which relays the sensory signal transmitting light to the retina.

Lab Activity 3 | Testing the Sensory and Somatic Motor Functions of the Face—Trigeminal Nerve: Cranial Nerve V

Cranial nerve V is a mixed nerve that emerges from the pons. It has three branches that provide somatic sensory information from the face. The mandibular branch also provides motor commands to the muscles of mastication.

1 | **Ophthalmic nerve (V1):** Sensations from the upper eyelid, eyebrow, nose, cornea, orbit, and skin of forehead.

2 | **Maxillary nerve (V2):** Sensations from the lower eyelid; upper lip, gums, teeth; cheek, nose; palate; parts of the pharynx.

3 | **Mandibular (V3):** Sensations from the skin of the jaw, lower gums, teeth, lips; palate; parts of the tongue; motor to muscles of mastication, including the masseter.

Materials Needed to Complete This Lab

- Cotton balls

A | Directions for Testing the Sensory Function of Each Branch of the Trigeminal Nerve

1 | Ask your lab partner to close their eyes.

2 | Using a soft item (such as a cotton ball), stroke their eyebrow, cheeks, and jaw.

3 | Repeat on both sides of the face.

4 | Record the results in Table 19.4.

Inability to sense this may indicate damage to the associated branch of the trigeminal nerve.

Table 19.4 | Sensory Functions of Cranial Nerve V

	Sensation Detected	Sensation Not Detected
Right Eyebrow		
Left Eyebrow		
Right Cheek		
Left Cheek		
Right Jaw		
Left Jaw		

B | Directions for Testing the Motor Function of the Trigeminal Nerve

1 | Ask your partner to clench their teeth together.

2 | Record the results in Table 19.5.

Inability to clench the teeth may indicate damage to the motor fibers of the mandibular branch of the trigeminal nerve.

Table 19.5 | Testing the Motor Function of the Trigeminal Nerve

Motor Function of Cranial Nerve V	Jaw Clench Normal	Jaw Clench Abnormal

Lab Activity 4 | Testing the Sensory and Somatic Motor Functions of the Face—Facial Nerve: Cranial Nerve VII

Cranial nerve VII is a mixed nerve that emerges from the pons. It has five branches that provide somatic sensory information from pressure receptors in the face, and special sensory information from the anterior two-thirds of the tongue about taste. Information about taste is received by the gustatory cortex in the frontal lobe. Cranial nerve VII also provides somatic motor control to muscles of the face and visceral motor control to lacrimal, nasal mucus, and salivary glands. The five branches of the facial nerve are the temporal, zygomatic, buccal, marginal mandibular, and cervical branches. Bell's palsy is a commonly known disorder that involves inflammation of the facial nerve, likely due to a viral infection. Signs and symptoms include loss of control of muscles of facial expression, loss of taste on the anterior two-thirds of the tongue, dry eyes, and dry mouth.

Materials Needed to Complete This Lab

- Lemon juice
- Salt water
- Sugar water
- Clean (sterile) Q-tips
- Gloves

A | Directions for Testing the Special Sensory Fibers of the Facial Nerve

1 | Ask your partner to close their eyes.

2 | You will dip a sterile Q-tip in one of the solutions provided in the lab (lemon juice, salt water, or sugar water).

3 | Give the Q-tip to your partner and ask them to touch the Q-tip to the **anterior** surface of their tongue.

4 | Ask them to identify the solution.

5 | Repeat with another solution.

6 | Record the results in Table 19.6.

Table 19.6 | Testing the Special Sensory Fibers of the Facial Nerve

	Correctly Identified	Incorrectly Identified
Solution 1:		
Solution 2:		

B | Directions for Testing the Somatic Sensory Fibers of the Facial Nerve

1 | Ask your partner to close their eyes.

2 | Use two gloved fingers to apply gentle pressure to the face near the cheek.

3 | Your partner will notify you when they feel the pressure.

4 | Repeat on the other side.

5 | Record the results in Table 19.7.

Table 19.7 | Testing the Somatic Sensory Fibers of the Facial Nerve

	Detected	Not Detected
Pressure on Face		

C | Directions for Testing the Somatic Motor Fibers of the Facial Nerve

1 | Ask your partner to perform the following actions:

a | Smile
b | Frown
c | Raise their eyebrows

2 | Watch for symmetry in both sides of the face as they perform each movement

3 | Record the results in Table 19.8.

Table 19.8 | Testing the Somatic Motor Fibers of the Facial Nerve

Facial Movement	Symmetric	Asymmetric
Smile		
Frown		
Raise Eyebrows		

D | Directions for Testing the Visceral Motor Fibers of the Facial Nerve

To assess the visceral motor fibers, inquire about dry mouth or dry eyes and record the results in Table 19.9.

Table 19.9 | Testing the Visceral Motor Fibers of the Facial Nerve

	Yes	No
Dry Mouth		
Dry Eyes		

Lab Activity 5 | Testing Hearing and Balance—Vestibulocochlear Nerve: Cranial Nerve VIII

Cranial nerve VIII, also known as the ***auditory nerve*** or ***acoustic nerve,*** emerges from the border of the pons and medulla oblongata. This nerve begins as two separate nerves, the *vestibular nerve* and the *cochlear nerve.* The vestibular nerve originates at the vestibule, a structure of the inner ear involved with the special sense of balance. The cochlear nerve originates at the cochlea, a structure of the inner ear involved with the special sense of hearing. Information about hearing is received by the auditory cortex located in the temporal lobe.

Materials Needed to Complete This Lab

- Tuning fork

A | Directions for Testing the Fibers of the Cochlear Portion of the Vestibulocochlear Nerve

1 | Your lab partner should be seated.

2 | Stand behind your lab partner and snap your fingers, one side at a time.

3 | Ask your partner to identify "right" or "left" as you do this.

4 | Record your results in Table 19.10.

Inability to hear the sounds may indicate damage to cranial nerve VIII.

Table 19.10 | Testing the Fibers of the Cochlear Portion of the Vestibulocochlear Nerve

	Detected	Not Detected
Finger Snap—Right Ear		
Finger Snap—Left Ear		

B | Weber and Rinne Tests

Next you will be conducting two standardized screening tests called Weber and Rinne Tests. These are commonly performed together clinically and are used to identify if one has a conductive or sensorineural deficit. ***Conductive hearing loss*** occurs when sounds cannot get through the outer and middle ear. This is often caused by an obstruction, such as ear wax or fluid from a cold, but may also be caused by damage to the outer ear, ear canal, ear drum, or ossicles in the middle ear. ***Sensorineural hearing loss*** occurs when there are nerve-related problems with the inner ear. Causes may include exposure to loud noises, head trauma, viral infections, or malformation of the inner ear. This type of hearing loss is also common with aging.

C | Directions for Testing the Fibers of the Cochlear Portion of the Vestibulocochlear Nerve Using the Weber Test

1 | The ***Weber test*** assumes that the patient has knowledge that they may have some hearing loss in one ear compared to another. Prior to beginning the test, ask your lab partner if they know if they have any loss of hearing in either ear.

2 | Your lab partner should be seated.

3 | Stand behind your lab partner and **gently** tap the tuning fork on the lab table surface, causing it to vibrate.

4 | Place the vibrating tuning fork on the top of your partner's head directly in midline.

5 | Ask if they hear the sound louder in one ear.

6 | If the sound is heard louder in the defective ear, they may have a conductive hearing loss in that ear.

7 | Record your results in Table 19.11.

In a normal Weber test, they should hear the sound equally in both ears. If the sound is heard louder in the normal ear, they may have a sensorineural loss in the defective ear.

Table 19.11 | Weber Test for Hearing Loss

	Right Ear	Left Ear
Is the person aware of any hearing loss? If so, indicate right or left ear.		
Normal Weber		
Conductive Hearing Loss		
Sensorineural Hearing Loss		

D | Directions for Testing the Fibers of the Cochlear Portion of the Vestibulocochlear Nerve Using the Rinne Test

1 | This test is typically performed after the Weber test, especially if the patient is unaware of any hearing loss.

2 | Your lab partner should be seated.

3 | Stand behind your lab partner and **gently** tap the tuning fork on the lab table surface, causing it to vibrate.

4 | Place the vibrating tuning fork on the mastoid process of the right ear until your partner tells you they no longer hear it.

5 | Remove the tuning fork from the mastoid process.

6 | Without striking it again, place the tuning fork just outside their right ear and ask them to tell you when they no longer hear the sound.

7 | Record your results in Table 19.12.

A normal Rinne test (positive) is when they still hear the sound when the tuning fork is placed outside their ear. This indicates that air conduction is equal or greater than bone conduction (AC > BC).

A negative Rinne test is when they do not hear the sound when the tuning fork is placed outside their ear. This indicates that bone conduction (BC) is greater than air conduction (AC) (BC > AC) and that there may be conductive hearing loss on that side.

Table 19.12 | Rinne Test for Hearing Loss

	Normal Rinne (Rinne +)	Abnormal Rinne (Rinne –)
Right Ear		
Left Ear		

E | Directions for Testing the Fibers of the Vestibular Portion of the Vestibulocochlear Nerve

To test the fibers of the vestibular portion of cranial nerve VIII, you will check their ability to maintain balance and equilibrium with changes to the orientation of the head. Here, you will assess position and acceleration.

Materials Needed to Complete This Lab

- Lab stool or chair with wheels

Lab Note: Individuals who are prone to motion sickness should not participate in this exercise as the subject (volunteer).

Procedures

1 | Adjust the lab stool so it is at the lowest height (for safety purposes).

2 | Ask for a volunteer to sit in the stool or chair and close their eyes.

3 | The volunteer should firmly grasp the edges of the chair to prevent falling from the chair.

4 | With the volunteer's eyes closed, cautiously move the chair in different directions (forward, backward, turned right, turned left) and ask the volunteer to identify the direction they were moved.

5 | Note if they could correctly detect the direction they were moved and document the results in Table 19.13.

6 | Repeat three times and note if there were any changes in their ability to accurately detect the direction moved.

Table 19.13 | Direction of Chair

	Attempt 1	Attempt 2	Attempt 3
Forward			
Backward			
Turn Right			
Turn Left			
Note Any Significant Differences			

F | Directions for Testing the Fibers of the Vestibular Portion of the Vestibulocochlear Nerve

This test of the vestibular portion of cranial nerve VIII checks for nystagmus during rotational head movements. ***Nystagmus*** is an involuntary rhythmic movement of the eyes. Movements can be side-to-side, up-and-down, or circular. There are many causes of nystagmus, with vestibular disorders being just one cause.

Materials Needed to Complete This Lab

- Lab stool or chair with wheels

Lab Note: Individuals who are prone to motion sickness should not participate in this exercise as the subject (volunteer).

Procedures

1 | Adjust the lab stool so it is at the lowest height (for safety purposes).

2 | Ask for a volunteer to sit in the stool or chair and close their eyes.

3 | The volunteer should firmly grasp the edges of the chair to prevent falling from the chair.

4 | With the volunteer's eyes closed, a lab partner will cautiously turn the chair to the left and right multiple times.

   a | Other students may need to hold the base of the lab stool and/or the volunteer to prevent the volunteer from falling.

5 | Stop the chair movement, have a lab partner grasp the volunteer's shoulders, then ask the volunteer to immediately open their eyes.

6 | Observe any abnormal movements of the eyes.

7 | Document your results in Table 19.14.

Table 19.14 | Nystagmus Test

	Detected	Not Detected
Right Eye		
Left Eye		

Lab Activity 6 | Testing for Taste and Swallowing—Glossopharyngeal Nerve: Cranial Nerve IX

Cranial nerve IX is a mixed nerve that emerges from the medulla oblongata. It contains somatic sensory fibers from the throat and soft palate, special sensory fibers from the posterior third of the tongue for the sense of taste, visceral sensory fibers that monitor blood pressure and gas concentrations in the carotid arteries, somatic motor fibers that control muscles used for swallowing, and visceral motor fibers that control the parotid salivary gland.

Materials Needed to Complete This Lab

- Lemon juice
- Salt water
- Sugar water
- Clean (sterile) Q-tips

A | Directions for Testing the Special Sensory Fibers of the Glossopharyngeal Nerve

1 | Ask your partner to close their eyes.

2 | You will dip a sterile Q-tip in one of the solutions provided in the lab (lemon juice, salt water, or sugar water).

3 | Give the Q-tip to your partner and ask them to touch the Q-tip to the **posterior** surface of their tongue.

4 | Ask them to identify the solution.

5 | Repeat with another solution.

6 | Record the results in Table 19.15.

Table 19.15 | Testing the Special Sensory Fibers of the Glossopharyngeal Nerve

	Correctly Identified	Incorrectly Identified
Solution 1:		
Solution 2:		

B | Directions for Testing the Somatic Motor Fibers of the Glossopharyngeal Nerve

1 | Ask your partner to swallow.

2 | Record the results in Table 19.16.

Table 19.16 | Testing the Somatic Motor Fibers of the Glossopharyngeal Nerve

	Normal	Abnormal
Swallowing		

Lab Activity 7 | Testing Heart Rate—Vagus Nerve: Cranial Nerve X

Cranial nerve X is another mixed nerve that arises from the medulla oblongata. The term "vagus" means wandering and earned this name because of its extensive branching through the body. It contains somatic sensory fibers for sensations from the external auditory meatus; special sensory fibers from taste receptors in the throat; visceral sensory information from organs such as the large intestine, respiratory tract, and esophagus; and visceral motor fibers controlling vital organs such as the heart and lungs, digestive organs, as well as muscles of the soft palate and throat. The vagus nerve carries parasympathetic fibers to the heart and major digestive organs, maintaining normal resting heart rate and digestive functions. This is why the parasympathetic division is often called "rest and digest."

Directions for Testing the Visceral Motor Function of the Vagus Nerve

1 | Your lab partner should be seated and relaxed.

2 | Using your second and third fingers, locate your partner's radial pulse.

3 | Measure their heart rate for 30 seconds.

4 | Multiply that number by two.

5 | Record the results in Table 19.17.

Normal resting heart rate should lie between 60–100 bpm.

Table 19.17 | Testing the Visceral Motor Function of the Vagus Nerve

	Normal	Abnormal
Calculated pulse in bpm:		

Lab Activity 8 | Testing Motor Functions—Accessory Nerve: Cranial Nerve XI

Cranial nerve XI, also know as the **spinal accessory** nerve, emerges from the medulla oblongata. It is composed of two branches, the *internal* and *external* branches, which both carry somatic motor fibers. The internal branch works with the vagus nerve to control muscles used for swallowing. The external branch controls the trapezius and sternocleidomastoid muscles.

Directions for Testing the External Branch of the Accessory Nerve

1 | Ask your partner to contract the sternocleidomastoid muscle by flexing the neck.

2 | Then, using your gloved hand, gently push against their forehead for resistance. This assesses the sternocleidomastoid bilaterally.

3 | Next, ask your partner to laterally flex their neck to the right then rotate to the left.

4 | Use your gloved hand to gently push against their forehead for resistance.

5 | Repeat on the left side, asking your partner to laterally flex to the left and rotate to the right.

6 | Test the trapezius muscle by asking your partner to elevate or "shrug" their shoulders.

7 | Apply resistance by gently pushing down on both shoulders.

8 | Record all results in Table 19.18.

Inability to perform the muscle action or weakness against resistance may indicate damage to the external branch of the accessory nerve.

Table 19.18 | Accessory Nerve Function Test

	Normal	Abnormal
Sternocleidomastoid *Bilateral—no resistance*		
Sternocleidomastoid *Bilateral—with resistance*		
Sternocleidomastoid *No resistance—right side*		
Sternocleidomastoid *No resistance—left side*		
Sternocleidomastoid *With resistance—right side*		
Sternocleidomastoid *With resistance—left side*		
Trapezius *No resistance—right side*		
Trapezius *No resistance—left side*		
Trapezius *With resistance—right side*		
Trapezius *With resistance—left side*		

Lab Activity 9 | Testing Motor Functions of the Tongue—Hypoglossal Nerve: Cranial Nerve XII

Cranial nerve XII emerges from the medulla oblongata and provides somatic motor control to the tongue.

Directions for Testing the Hypoglossal Nerve

1 | Ask your lab partner to stick out their tongue.

2 | Record the results in Table 19.19.

3 | If the tongue veers to one side, this may indicate damage to the hypoglossal nerve on that side.

Table 19.19 | Testing Motor Functions of the Tongue

	Normal	Veers Right	Veers Left
Tongue Position			

Laboratory 19 Review Questions

Cranial Nerve Testing

Name: ____

1 | List the general and special senses.

2 | Explain how the optic nerve is involved in the testing of the pupillary response to light and the ability to follow a moving penlight during the evaluation of cranial nerves III, IV, and VI.

3 | The ability to swallow is dependent on normal function of cranial nerve __________ and __________, as well as the ____________________ branch of the accessory nerve.

4 | The gag reflex is coordinated by cranial nerves ____________________ and ____________________.

5 | During the consensual light reflex, if a light shines into the right eye, the right pupil will ________________ *(constrict/dilate/remain unchanged)* while the left pupil will ________________ *(constrict/dilate/remain unchanged).*

6 | Which lobe of the brain are each of the following located?

a | Gustatory cortex ______________________________

b | Olfactory cortex ______________________________

c | Visual cortex ______________________________

d | Auditory cortex ______________________________

7 | Which cranial nerves have no direct function with general or special senses?

8 | Dry mouth may be due to damage of cranial nerve ______________ or cranial nerve ______________.

9 | Autonomic functions are provided by which type of neuron? ________________________.

10 | Autonomic functions are dependent on information coming from which type of neuron? ____________

11 | Which two cranial nerves do not arise from the brain stem?

12 | Which cranial nerve provides parasympathetic control of the heart?

13 | CN VIII is often referred to as the *acoustic* or *auditory* nerve. Explain why these may not be the most accurate name for this nerve.

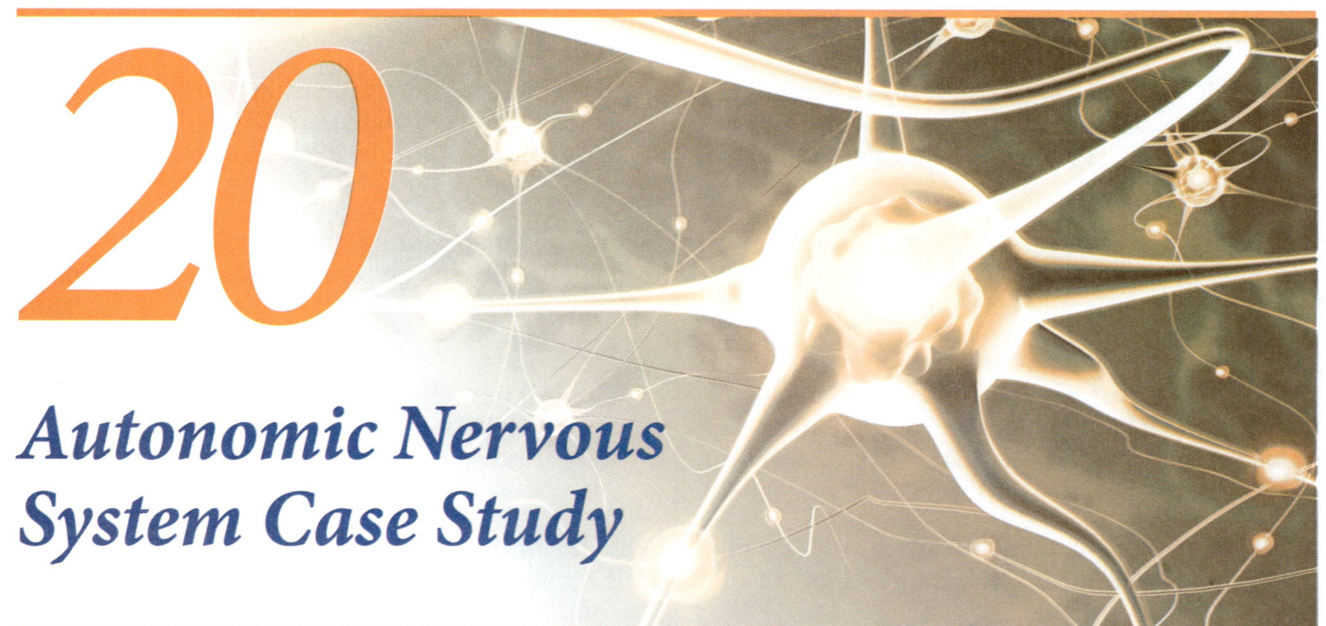

20 Autonomic Nervous System Case Study

Introduction

The autonomic nervous system (ANS) consists of both the parasympathetic and sympathetic divisions. The ANS regulates many visceral reflexes and helps to maintain the homeostasis of organ systems. Most organs are innervated by both divisions. However, the response(s) at the effectors (target organs) tend to be contrasting. For example, if the sympathetic system speeds up the heart rate, the parasympathetic system brings the heart rate back down to the resting state.

The ANS efferent pathways to the organs involve a two-neuron pathway. The first neuron is known as the **preganglionic neuron** and it originates in the central nervous system. The second neuron in the two-neuron pathway is the **postganglionic neuron** (also known as a ganglionic neuron). It originates in an autonomic ganglion, located outside of the CNS, and terminates at the effector (organ). There is a difference in the type of neurotransmitters released by the postganglionic neurons. The parasympathetic postganglionic neurons (fibers) release acetylcholine (Ach) as their neurotransmitters; these fibers are referred to as ***cholinergic.*** The sympathetic postganglionic neurons (fibers) primarily release norepinephrine (**NE**) as their neurotransmitters; these fibers are referred to as ***adrenergic.***

In order for effectors to respond to either the sympathetic or parasympathetic divisions, the cells in those organs must contain receptors for the specific ANS neurotransmitters. The cells which respond to Ach need to have ***cholinergic receptors*** that bind to Ach. Any cell that responds to NE must have ***adrenergic receptors***. Additionally, there are subcategories of both cholinergic and adrenergic receptors. Cholinergic receptors are classified as either **nicotinic** or **muscarinic.** When Ach binds to nicotinic receptors, an excitatory postsynaptic potential (EPSP) occurs; however, binding to muscarinic receptors can be either excitatory or inhibitory, dependent upon the specific effector. Adrenergic receptors have two major categories, alpha (α) receptors and beta (β) receptors. There are additional subcategories of the α and β receptors: α_1, α_2, β_1, and β_2. Dependent upon the specific adrenergic receptor, the effects could be either excitatory or inhibitory. Many medications target specific receptors by either imitating the actual neurotransmitter (**mimetic** drugs) or blocking the receptors (blocking drugs), preventing the ability of the actual neurotransmitter from binding to the receptors.

Materials Needed for This Lab

- Textbook
- Applications manual

Case Study

While shopping at Target, 56-year-old John and his wife, Kathy, stopped by the pharmacy to have their free blood pressure (BP) reading done. Kathy's blood pressure was 125/76 mm Hg, and John's BP was 176/94 mm Hg. Seeing these values and noting the information on the BP machine, John dutifully made an appointment to see his physician.

At the first appointment, Dr. Martin did a thorough check up on John, including labs. She noted that his BP in the office was 170/90 mm Hg. After consulting with John about his diet and exercise habits, she recommended that he start a program of exercise and begin a reduced fat diet. An appointment for two months in the future was made.

At the next appointment, John had lost eight pounds and was walking regularly. However, his blood pressure was still elevated at 172/92 mm Hg. Dr. Martin recommended that John stay on his diet and exercise plan, but added a blood pressure medication called Minipress (prazosin) to the regimen. This medication is an α_1 blocker. An appointment was made for the following month.

At this appointment, John's BP was now 142/82 mm Hg. Dr. Martin noted that his BP was coming down nicely. In addition to the medication, John had lost an additional three pounds and had continued his walking.

If you were part of Dr. Martin's medical team, think through the reasoning for prescribing this medication.

Answer the following questions, using the information above as well as your textbook (locate the tables in the ANS chapter to assist you in answering these questions) and Applications Manual (nervous system chapter).

John's Medical Status

1 | List some of the causes of high blood pressure.

2 | Which of these factors does John appear to have?

3 | How ***specifically*** would changing diet and exercise help in reducing John's blood pressure?

4 | Can you think of other ways John could reduce his high blood pressure?

ANS Questions

5 | What is a sympathomimetic drug? How does it work on the body?

6 | What is a parasympathomimetic drug? How does it work on the body?

7 | Are alpha receptors part of the sympathetic or parasympathetic division?

8 | Are alpha receptors cholinergic or adrenergic? ______________________

9 | By using your textbook, fill in the table below. List the two categories of **alpha receptors.** Describe their locations (body structure) and their normal sympathetic effect(s) at each location. (We are not considering parasympathetic effects.)

Table 20.1 | Types of Alpha Receptors

Receptor Type	Body Structure	Normal Effect (Sympathetic)

10 | Are beta receptors part of the sympathetic or parasympathetic division? ______________________

11 | Are beta receptors cholinergic or adrenergic? ____________________

12 | By using your textbook, fill in the table below. List the two categories of **beta receptors.** Describe their locations (body structure) and their normal sympathetic effect(s) at each location. (We are not considering parasympathetic effects.)

Table 20.2 | Types of Beta Receptors

Receptor Type	Body Structure	Normal Effect (Sympathetic)

13 | Minipress (prazosin) is a sympathetic blocking agent. Explain this term to John. ____________________

____

____

14 | How ***specifically*** will this medication help reduce John's blood pressure?

15 | How does a "blocking agent" work?

16 | Will the Minipress that John is taking affect other tissues or organs? Why or why not?

Cardiac Output: One factor that can affect blood pressure is the **cardiac output (CO).** This is the amount of blood leaving the heart each minute. There are two factors that play a role in CO—the volume of blood leaving the heart (also known as the **stroke volume**) and the number of beats per minute (the **heart rate (HR)**). Affecting either of these two factors (stroke volume or heart rate) can affect the blood pressure.

17 | Explain the effect that stimulation of a β_2 receptor would have on cardiac muscle cells.

a | How would that affect the heart rate (HR) and blood pressure?

b | What would the effect be on bronchial smooth muscle cells?

18 | Explain the effect that blocking β_1 receptors would have on cardiac muscle cells.

a | How would that affect the HR and blood pressure?

b | What would the effect be on bronchial smooth muscle cells?

19 | Explain the effect of stimulating α_1 receptors on cardiac muscle tissue.

a | How would that affect the HR and blood pressure?

b | What would the effect be on bronchial smooth muscle cells?

20 | Using Tables 20.1, 20.2, and your Applications Manual, can you think of other types of medications (blockers or mimetics) that John could use to reduce his high blood pressure?

21 Endocrine Case Studies

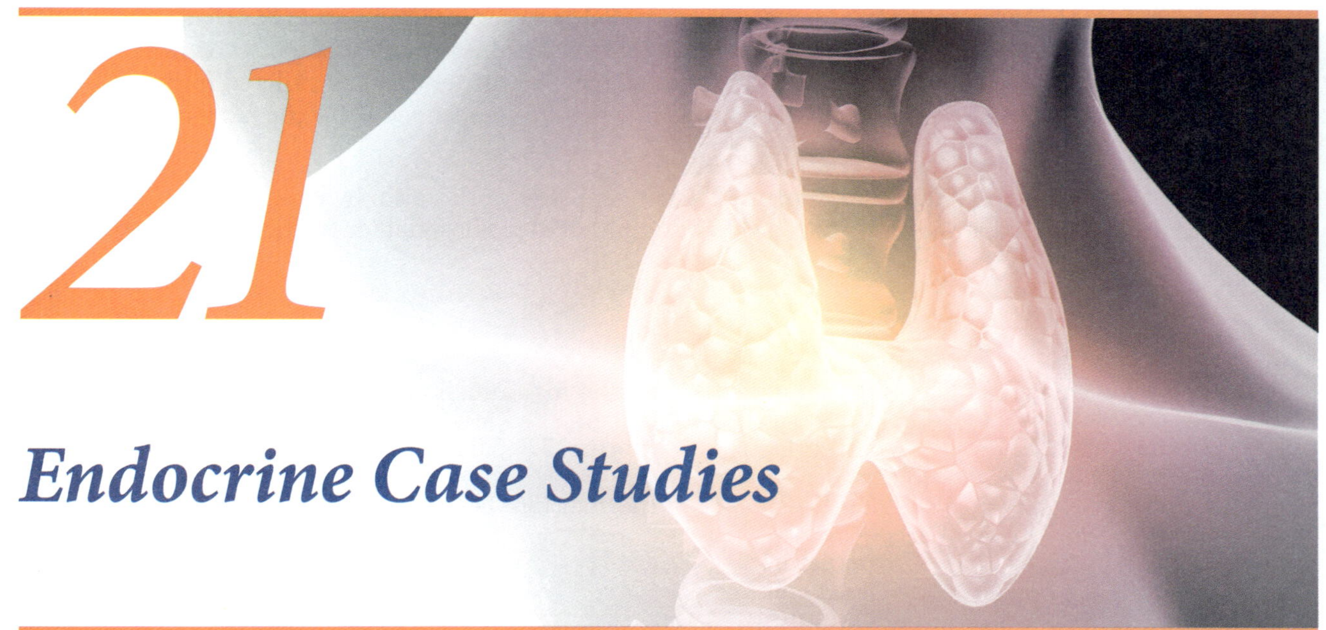

Materials Needed to Complete This Lab

- Textbook
- Applications manual

Instructions

- Your instructor will assign you to one of these case studies.
- Based on instructor preference, you may be assigned to work on your case study as a group or individually.
- Reference your textbook and applications manual to assist you in completing this lab.

Case Study 1 | Addison's Disease

Patient History

Jamal, a 54-year-old male, was brought to the emergency room by ambulance after being found collapsed on the floor of his home. The paramedics called the ER with the following information:

Blood pressure	80/65 mm Hg
Pulse	140 beats/minute
Blood glucose	55 mg/dl
Physical findings	Patient appears dehydrated and weak
Past medical history	Past history of malignant kidney tumor
	Recent history of exploratory abdominal surgery in connection to the tumor finding

Upon arrival at the emergency room, Jamal was examined and the physician noted abnormal pigmentation of the abdominal scar. His wife reported that Jamal had a recent history of nausea and vomiting as well as weight loss accompanied by muscle weakness and fatigue. She thought he had suffered from the flu. She recalled that there were times when he got up suddenly that he complained of feeling faint and had a rapid heartbeat.

Jamal was revived with IV fluids but remained lethargic. Blood work was drawn and sent to the STAT lab. The results indicated that the patient needed to be admitted to the hospital for further testing.

Lab Results

The following are Jamal's lab results from the ER and after his admission to the hospital:

- Serum K^+ — 6.0 mEq/L (normal 3.8–5.0 mEq/L)
- Serum Na^+ — 122 mEq/L (normal 136–142 mEq/L)
- Urine Na^+ — hypernaturic
- ECG — arrhythmia
- Arterial blood gases — metabolic acidosis
- Aldosterone — decreased levels
- Cortisol — decreased levels
- MSH — increased levels
- ACTH — increased levels

Use the information above, as well as your textbook and Applications Manual to answer the following questions:

1 | What endocrine gland(s) are involved?

2 | What is the normal feedback loop of the gland(s) and organ(s) involved? Develop this feedback loop.

3 | Where did the feedback loop fail that resulted in the disorder and symptoms? Develop the diseased loop.

4 | What should be occurring at the cellular level of this patient **if** his body system(s) were working properly?

5 | What is occurring at the cellular level of this patient due to the disease?

6 | How are the disease symptoms and the laboratory findings related?

7 | Is this a primary or secondary disease?

8 | What are the treatment options (might include pharmacological, surgical, or lifestyle changes)? Provide specific detail as to how each of the selected treatments would provide beneficial results. If discussing pharmacological treatments, be sure to explain specifically how the drugs would act on the body.

Reference Lab Tests and Values

From The Merck Manual, 13th edition.

Adrenocorticotropic Hormone (ACTH) (plasma)

This pituitary hormone may be measured by radio immunoassay (RIA). Its presence distinguishes primary adrenal tumors from adrenal hyperplasia and may help in establishing the existence of pituitary tumors. Markedly elevated plasma concentrations are present in ectopically produced ACTH. Changes may be used as a guide to effective tumor ablation.

Aldosterone (plasma, serum, or urine)

This is a potent mineralocorticoid produced by the adrenal gland which, in excess, causes hypokalemic alkalosis, sodium retention, and hypertension. It is measured by RIA. Measurement may be useful in establishing the existence of aldosterone-producing tumors or in assessing the degree of secondary hyperaldosteronism that may occur in cyclic edema, nephritic syndrome, cirrhosis of the liver with ascites, and cardiac failure.

Cortisol (plasma)

Plasma cortisol measurement is useful for evaluating adrenal cortical function and distinguishing hyperadrenalism. It is best measured before and then eight hours after stimulation with IV ACTH. Plasma cortisol levels are determined by RIA.

Sodium (Na^+) (serum)

Hyponatremia can be caused by excessive use of diuretics, inappropriate ADH secretion, diarrhea, untreated diabetes, excessive sweating and adrenal insufficiency. However, hyponatremia usually reflects an excess of water in the body rather than a low sodium level. Serum sodium levels are elevated in dehydration, uremia, and aldosteronism. The determination is usually made by flame photometry but may be made by ion-specific electrodes.

Potassium (K^+) (serum)

Nearly 90% of total body K^+ is intracellular and about 8% is in the bone; less than 1% is present in serum. Serum K^+ levels are decreased due to the following: by certain diuretics; by use of K^+-free parenteral solutions; by GI losses due to vomiting, diarrhea, or gastric, intestinal or biliary drainage; and during attacks of hypokalemic periodic paralysis. K^+ levels are sometimes lowered in primary aldosteronism. They are elevated in uncontrolled diabetes mellitus, acute and certain phases of chronic renal disease, Addison's disease, adrenal insufficiency, and respiratory distress syndrome in the newborn. The determination may be made by the use of an ion-specific electrode.

Case Study 2 | Hypothyroidism

Patient History

Marissa, a 38-year-old female, scheduled an appointment with her family physician because of a variety of symptoms that have appeared gradually over the past several months. She informed the doctor that she has been feeling very tired and is intolerant of the cold. She also complained of dry, scaly skin and thinning, brittle hair. She has puffiness in her face and is experiencing constipation. Normally very active, now she has a general lack of interest in any activity. Her husband and children have also noticed a change in her personality.

The physician examined her and ordered a series of laboratory tests. The results of the tests are below:

Test Results

- Heart rate 54 beats/minute
- Blood pressure 85/68 mm Hg
- Hemoglobin 9.4 mg/dl (normal 15.0 mg/dl)
- T_3 Total = 80 ng/dl
  Free = 0.1 ng/dl
- T_4 Total = 3.0 ug/dl
  Free = 0.6 ng/dl
- TSH elevated as tested by RIA
- Radiology showed an enlarged heart.

Reference Lab Tests Normal Values

(These values can vary slightly dependent upon resource cited.)

- ***Total T_4 (Serum)*** normal value = 4.5–12.5 ug/dl
- ***Free T_4 (Serum)*** normal value = 0.7–2.0 ng/dl
- ***Total T_3 (Serum)*** normal value = 80–220 ng/dl
- ***Free T_3 (Serum)*** normal value = 0.2–0.5 ng/dl

Use the information above, as well as your textbook and Applications Manual to answer the following questions:

1 | What endocrine gland(s) are involved?

2 | What is the normal feedback loop of the gland(s) and organ(s) involved? Develop this feedback loop.

3 | Where did the feedback loop fail that resulted in the disorder and symptoms? Develop the diseased loop.

4 | What should be occurring at the cellular level of this patient ***if*** her body system(s) were working properly?

5 | What is occurring at the cellular level of this patient due to the disease?

6 | How are the disease symptoms and the laboratory findings related?

7 | Is this a primary or secondary disease?

8 | What are the treatment options for Marissa? These might include pharmacological, surgical, or lifestyle changes. Provide specific detail as to how each of the selected treatments would provide beneficial results. If discussing pharmacological treatments, be sure to explain specifically how the drugs would act on the body.

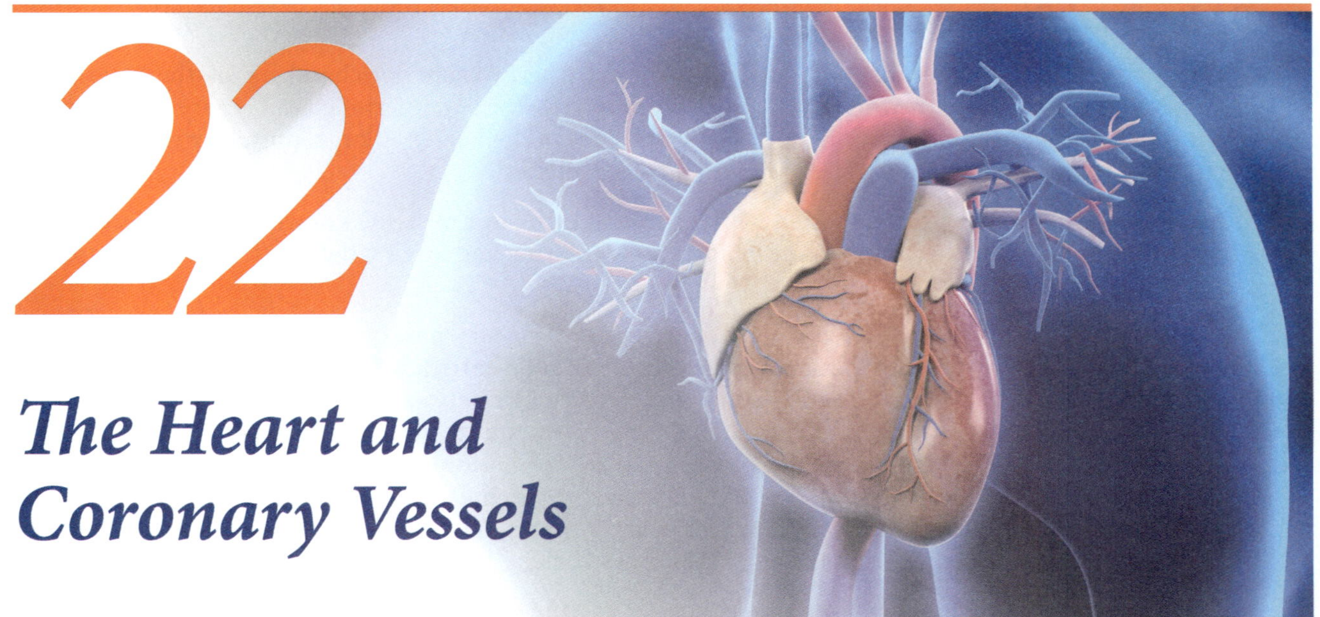

22 The Heart and Coronary Vessels

A | The Heart Review

A | External Features

The heart is a muscular pump located in the thoracic cavity. It is surrounded by a tough sac, the ***fibrous pericardium,*** which serves to anchor the heart in place, protect it, and prevent it from overfilling with blood.

The inner portion of the sac, the ***serous pericardium,*** is thinner and not as strong. It consists of two portions: the ***parietal layer,*** which is fused to the fibrous pericardium, and the ***visceral layer,*** which is also known as the ***epicardium.*** The epicardium lies directly on top of the heart and appears as a slippery, shiny membrane on the exterior surface of the heart.

The broad, upper portion of the heart is known as the ***base,*** and the inferior tapered end is known as the ***apex.*** The "dog ear" looking flaps on the upper surface of the heart are the ***auricles.*** They serve to increase the surface area of the chambers (atria) that lie below them internally.

An external groove on the anterior surface (front) of the heart is known as the ***anterior interventricular sulcus.*** It divides the two lower chambers of the heart (ventricles) externally.

Also associated with the heart are several large vessels, which transport blood into and out of the chambers of the heart. Entering the right side of the heart (right atrium) are the ***superior vena cava*** (transporting deoxygenated blood from the upper body) and ***inferior vena cava*** (transporting deoxygenated blood from the lower body). Leaving the right side of the heart from the right ventricle is the ***pulmonary trunk*** which divides into two ***pulmonary arteries.*** These carry deoxygenated blood to the lungs. Entering the left side of the heart (left atrium) are the four ***pulmonary veins,*** which deliver oxygenated blood from the lungs back to the heart. The large ***aorta*** carries oxygenated blood from the left side of the heart (left ventricle) to the entire body.

B | Internal Features

Internally there are four chambers (cavities) in the heart. The two superior chambers are the ***atria (right atrium*** and ***left atrium,*** on their respective sides of the heart). The two inferior chambers are the ***ventricles (right ventricle*** and ***left ventricle,*** on their respective sides of the heart). A large, muscular wall known as the ***interventricular septum*** separates the two ventricles internally.

The four chambers of the heart are lined with a smooth, thin layer of tissue known as the ***endocardium,*** which is continuous with the lining of the major vessels entering and leaving the heart. (The valves are also composed of this tissue.)

There are four valves in the heart, which help to ensure the one-way flow of blood through the heart. The ***atrioventricular (AV) valves*** are located between the atria and ventricles. The AV valve between the right atrium and right ventricle is the ***tricuspid valve,*** and the AV valve between the left atrium and left ventricle is the ***bicuspid*** or ***mitral valve.*** On the bottom of these valves are string-like cords known as the ***chordae tendineae.*** They are attached to the cusps ("bumps") on the bottom of the valves and to the ***papillary muscles*** on the walls of the ventricles. The papillary muscles are muscular columns that contract and pull on the chordae tendineae to keep them taut; this ensures that the AV valves will close tightly and prevent blood from flowing backwards from the ventricles into the atria.

The ***semilunar valves*** are found between the ventricles and the large arteries which exit from the ventricles. The semilunar valve exiting the right ventricle to the pulmonary trunk is the ***pulmonary semilunar valve.*** The semilunar valve exiting the left ventricle to the aorta is the ***aortic semilunar valve.*** The semilunar valves have no tendon cords and have their cusps attached directly to the inside walls of the large arteries exiting the ventricles.

Lab Note: Remember that the purpose of valves is to ensure that blood flows unidirectionally. The AV valves allow blood to flow from the atria to the ventricles. The semilunar valves allow blood to flow from the ventricles into the arteries that carry the blood out of the heart.

The atria and ventricles are composed of the ***myocardium*** which is the muscular layer of the heart. It is responsible for contracting and pumping blood through the heart, the lungs, and the body. The walls of the atria are thinner because they do little active pumping; gravity allows most of the blood to flow from the atria (upper chambers) to the ventricles (lower chambers). In contrast, the ventricular myocardium is thicker as they must pump blood out of the heart. Of the two ventricles, the left ventricle has the thickest walls because it must contract with enough force to pump blood through the entire body. ***You can use this information to help you determine the heart chambers in the preserved hearts.***

C | Coronary Vessels: The Blood Supply to the Heart

Even though there is a flow of blood *through* the chambers of the heart at all times, the myocardium (heart muscle) must have its own constant supply of oxygenated blood in order to pump effectively. This oxygenated blood is furnished by the ***coronary arteries,*** which branch directly from the root of the aorta as it leaves the left ventricle. Therefore, the coronary arteries get a fresh supply of oxygenated blood with each contraction of the left ventricle.

The ***right coronary artery*** lies underneath the right auricle and travels around to the dorsal surface of the heart. It supplies the right atrium and parts of both ventricles with oxygenated blood.

The right coronary artery gives rise to the ***marginal branch*** which travels along the underside of the right ventricle. The right coronary artery continues along the dorsal surface (between the ventricles) as the ***posterior interventricular branch*** or ***artery*** (some call this the posterior descending artery), which supplies the interventricular septum with blood.

The ***left coronary artery*** supplies the left ventricle and left atrium. It divides to form two branches. It divides near the left atrium and travels under the left auricle and around to the dorsal surface of the heart as the ***circumflex branch*** (or ***artery***) which eventually meets up with the right coronary artery branches. The other branch of the left coronary artery follows the anterior interventricular sulcus and is known as the ***anterior interventricular artery or left anterior descending (LAD) artery.***

Please Note: All coronary arteries have the capability to "bud" off new arteries called ***collateral vessels.*** These can be stimulated to develop via aerobic exercise that leads to cardiovascular fitness. As the collateral vessels grow across the surface of the heart, they often will join together to form ***anastomoses*** or connections. These anastomoses are beneficial in providing continual blood flow to the myocardium.

Blockage of any of the coronary arteries can interfere with blood flow to the myocardium and, therefore, with heart function. This can result in a ***myocardial infarction*** or ***heart attack.***

Once the myocardium has received its supply of oxygen, the cardiac veins collect the deoxygenated blood produced by the cardiac muscle cells.

The ***great cardiac vein*** collects deoxygenated blood from the front of the heart. It is located in the anterior interventricular sulcus. The ***small cardiac vein*** is a very small vessel lying underneath the right auricle, and it collects from this area of the heart. The ***middle cardiac vein*** collects deoxygenated blood from the posterior surface of the heart. The middle cardiac vein lies near the posterior interventricular artery.

All collected venous blood is then dumped into the ***coronary sinus,*** a large vein found on the dorsal side of the heart. Once the blood has been collected here, it is delivered into the right atrium of the heart, where it will then be pumped back to the lungs for more oxygen.

D | Blood Flow through the Heart

Blood flows into both atria and out of both ventricles at the same time. However, for ease of explanation, most texts and instructors begin at the right side of the heart and finish with the left side. **Keep this in mind!!**

Right Side

The right side of the heart receives and pumps deoxygenated blood (high in CO_2 and low in O_2). The blood is received from the body, pumped to and through the lungs where it can pick up oxygen and release the carbon dioxide and then is returned to the heart.

→ From body systems → superior vena cava (from upper body), inferior vena cava (from lower body), coronary sinus (from heart muscle) → right atrium → tricuspid valve → right ventricle → pulmonary semilunar valve → pulmonary trunk → pulmonary arteries → to lungs.

Left Side

The left side of the heart receives oxygenated blood from the lungs. After traveling through the left side of the heart, this blood is pumped to all systems of the body to be used in chemical reactions such as the production of ATP (energy).

→ From lungs → pulmonary veins → left atrium → bicuspid (mitral) valve → left ventricle → aortic semilunar valve → aorta → body systems.

Lab Activity

A | The Heart Model

Using the heart models, the notes above, and other references, please locate the following structures.

1 | External Features

- Apex
- Base
- Anterior interventricular sulcus
- Auricle
- Epicardium

Great Vessels of the Heart

- Superior vena cava
- Inferior vena cava
- Pulmonary trunk
- Pulmonary arteries
- Pulmonary veins
- Aorta

Lab Tip: On the heart models, the arteries are colored red (to indicate oxygenated blood) and the veins are colored blue (to indicated deoxygenated blood). The exceptions to this color "rule" are the pulmonary arteries and pulmonary veins. ***The pulmonary arteries are BLUE*** because they carry deoxygenated blood and the ***pulmonary veins are RED*** because they carry oxygenated blood.

2 | Internal Features

Chambers and Valves

- Right atrium
- Left atrium
- Right ventricle
- Left ventricle
- Tricuspid valve
- Bicuspid (mitral) valve
- Pulmonary semilunar valve
- Aortic semilunar valve

Other Internal Structures

- Interventricular septum
- Chordae tendineae
- Papillary muscles
- Endocardium
- Myocardium

3 | Coronary Vessels

Arteries (colored red)

- Right coronary artery
- Left coronary artery
- Circumflex artery
- Anterior interventricular artery (left anterior descending)
- Posterior interventricular artery
- Marginal artery

Veins (colored blue)

- Great cardiac vein
- Middle cardiac vein
- Small cardiac vein
- Coronary sinus

B | The Preserved Heart

Materials Needed to Complete this Lab

- Pre-dissected preserved heart (sheep or pig)
- Dissecting tray
- Scalpel (if not already dissected)
- Probes
- Disposable gloves

Procedure

1 | Working in groups of four, obtain a preserved heart from your instructor and place upon a dissecting tray. (The hearts may be either from sheep or pigs, depending upon availability.)

2 | ***If the heart has been previously dissected,*** orient yourself by locating the ***left side*** of the heart. You may locate the left side by examining the thickness of the chambers of the heart. Recall that the ventricular walls are much thicker than the atrial walls. The walls of the ***left ventricle*** are the thickest! First locate the interventricular septum (the wall of myocardium between the ventricles internally) and then look to the *outer* walls of the ventricles. The wall which is the thickest is the left ventricle.

If the heart has not been previously dissected, you may use your scalpel to divide the heart almost into two pieces. To make a frontal section of the heart, start cutting at the top right side of the heart. Cut down around the ventricles until you end up at the top of the other side. Don't sever the very top of the heart. Now you may open the heart from the bottom up.

3 | Locate the following features on the preserved heart:

a | **External Features**

- Apex
- Base
- Anterior interventricular sulcus
- Auricle
- Epicardium

Lab Note: The anterior interventricular sulcus is often filled with fat on the preserved heart so may be difficult to locate. In addition, the heart may have become misshapen during processing and shipping so anterior and posterior surfaces may be difficult to identify.

Lab Note: You may be able to locate some of the great vessels of the heart, but they are often difficult to identify due to the processing of the hearts. Because of the difficulty identifying many of these great vessels, you will not be tested on these structures on the preserved hearts.

b | **Internal Features**

Chambers and Valves

- Right atrium
- Left atrium
- Right ventricle
- Left ventricle
- Tricuspid valve
- Bicuspid (mitral) valve

Other Internal Structures

- Interventricular septum
- Chordae tendineae
- Papillary muscles
- Endocardium
- Myocardium

Lab Note: Dependent upon the quality of dissection of the preserved heart, you may not be able to locate the semilunar valves. So you will not be tested on those features on the preserved hearts.

In addition, you will not be tested on the coronary vessels on the preserved hearts.

4 | When finished with the preserved hearts, return them to their containers. Wash and dry all dissecting materials.

C | The Human (Cadaver) Heart (if available)

Materials Needed to Complete this Lab

- Preserved human heart
- Dissecting tray
- Probes
- Disposable gloves

Procedure

If available, locate the same structures on the human heart as you found on the preserved animal heart. You may also be able to observe the ***pericardial sac*** on the cadaver heart.

Your instructor may choose to perform this section of the lab as a demonstration.

Required Structures of the Heart

After completion of the lab, you should be able to identify and correctly spell the following heart structures on the preserved/dissected hearts and heart model.

Structures on the Preserved Heart

1 | Apex

2 | Atrium (right and left)

3 | Auricle

4 | Base

5 | Bicuspid (mitral) valve

6 | Chordae tendineae

7 | Endocardium

8 | Epicardium

9 | Interventricular septum

10 | Myocardium

11 | Papillary muscle

12 | Tricuspid valve

13 | Ventricle (right and left)

Structures on the Heart Model

Great Vessels of the Heart

1 | Aorta

2 | Inferior vena cava

3 | Pulmonary trunk

4 | Pulmonary artery

5 | Pulmonary vein

6 | Superior vena cava

Coronary Vessels of the Heart

Arteries

1 | Anterior interventricular artery (left anterior descending)

2 | Circumflex artery

3 | Left coronary artery

4 | Marginal artery

5 | Posterior interventricular artery

6 | Right coronary artery

Veins

1 | Coronary sinus

2 | Great cardiac vein

3 | Middle cardiac vein

4 | Small cardiac vein

Other structures

1 | Anterior interventricular sulcus

2 | Aortic semilunar valve

3 | Pulmonary semilunar valve

Laboratory 22 Review Questions

The Heart and Coronary Vessels

Name: ______________________________

1 | List the three layers of the heart wall.

a |

b |

c |

2 | Blood flowing from the venae cavae will enter the ______________________________.

3 | Blood leaving the right atrium will pass through which valve? ______________________________

4 | Blood leaving the left atrium will pass through which valve? ______________________________

5 | Blood returns to the heart from the lungs through which vessels? ______________________________

6 | Blood leaving the right ventricle would travel through the ______________________________ valve and then on to the lungs through which vessels? ______________________________

7 | The aortic semilunar valve is found between the ______________________________ and the ______________________________.

8 | Which chamber of the heart has the thickest myocardium and why?

9 | All blood from the coronary veins empties into the ______________________________.

10 | Where do the right and left coronary arteries originate? ______________________________

Match the following:

11 | ___________	groove on the anterior heart surface	*a* |	chordae tendineae
12 | ___________	left AV valve	*b* |	papillary muscles
13 | ___________	cords attaching valve to heart wall	*c* |	interventricular sulcus
14 | ___________	projection in ventricles for attachment of valve cords	*d* |	endocardium
15 | ___________	innermost wall of heart; lining	*e* |	bicuspid

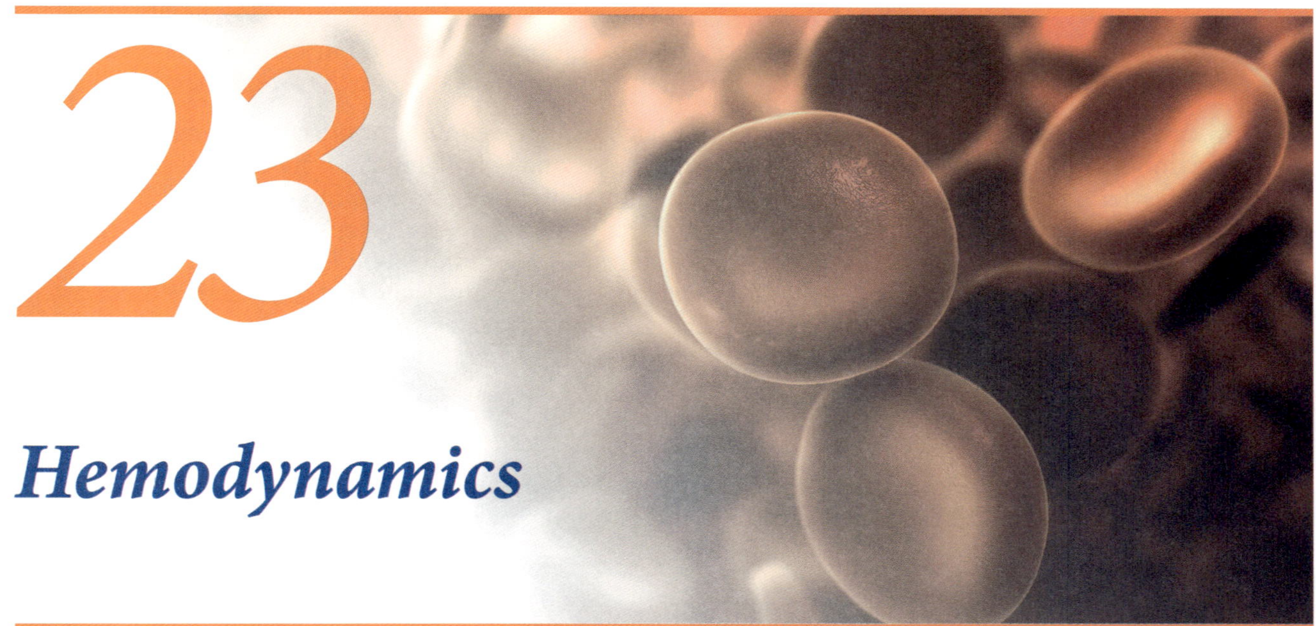

23 Hemodynamics

Introduction to Hemodynamics

Relationship between Blood Flow, Pressure, and Resistance

Blood flow is the movement of blood through a blood vessel. Blood flow through a blood vessel or a series of blood vessels is decided by two factors: the first is the pressure difference between the two ends of the vessel (where blood comes in and where blood exits) and the second is the **resistance** of the vessel to blood flow. The difference in pressure is the driving force for blood flow, and resistance is an inhibition to flow. Resistance is generated by the force called **friction.** Friction is created by the contact of the blood components (cells, water, proteins) with the inner lining of the blood vessel (endothelium). The more contact between blood and the blood vessel lining the greater the force of friction is. Therefore, resistance is greater.

The measure of **blood flow (*F*)** is related to the size of the pressure difference (ΔP) or **pressure gradient.** The *direction* of blood flow is decided by the direction of the pressure gradient and always is from ***high to low pressure.*** For example, during ventricular ejection, blood flows from the left ventricle into the aorta and not in the other direction because pressure in the ventricle is higher than pressure in the aorta. This is different from **blood pressure** which is the measurement of the blood pushing against the vessel wall.

How Resistance to Blood Flow is Calculated

As described above, friction between the blood and blood vessel lining is what generates resistance to blood flow. The relationship between resistance, blood vessel diameter (or radius), and blood viscosity is described by the **Poiseuille equation. *R* = Resistance, η = viscosity, *l* = length, *r* = radius.**

$$R = \frac{8\eta l}{\pi r^4}$$

Pressure and Resistance

As we have learned in lab and lecture anatomy review, there are different vessel types in the body. Large arteries decrease to arterioles, arterioles lead to capillaries, capillaries empty into venules, and venules increase to larger veins. When we consider pressure as we move to the capillaries, we see a decrease in pressure. However, a decrease in the radius of a vessel causes an increase in resistance. The pressure decreases as the blood approaches the capillaries because they are the smallest in size. Yet, they are not the vessels with the greatest resistance; the vessels with the greatest resistance are the arterioles.

How can this be explained? To understand this phenomenon, we must consider two things. The first is where do we see the greatest drop in pressure and the second is how much of the body does that vessel type cover (cross sectional area of the body). Even though we have far more capillaries and they cover a greater part of the body than any other blood vessel type, they are not the vessels that generate the greatest resistance. It is the arterioles that generate the greatest resistance and why they are called the resistance vessels. This is because arterioles cover a sizable part of the body, and their **vasoconstriction** (decrease in vessel radius) and **vasodilation** (increase in vessel radius) have a greater impact on resistance than the other vessels in the body.

Lab Note: Although units of measurement are important in equations, you will not be required to know the units used in the calculations you perform in this lab.

Viscosity and Resistance

The thickness of a fluid is called **viscosity.** The more viscosity a fluid has, the greater the resistance it has when flowing through a tube. We can make the connection to resistance in the following way: ***The greater the viscosity, the greater the resistance.***

The blood's viscosity is created by the presence of the plasma proteins and the formed elements (erythrocytes—red blood cells, leukocytes—white blood cells, and thrombocytes—platelets). **Albumin,** a plasma protein created by the liver, is the greatest contributor to the viscosity of blood. When comparing blood viscosity to water viscosity (viscosity of 1), blood is 5 times thicker/more viscous than water. Blood viscosity averages between 3–5.

When comparing vessel length, radius, and viscosity, viscosity has the smallest impact on resistance.

Pre-Lab Problems

1 | A man suffers a stroke caused by partial occlusion/blockage of his left internal carotid artery. An evaluation of the carotid artery using magnetic resonance imaging (MRI) shows a 75% reduction in its radius. Normally, the radius is 1 cm. His blood has a viscosity of 4.0. The length of his internal carotid is 15 cm.

a | Using Poiseuille's equation, calculate what the resistance would be in an unblocked internal carotid.

b | Calculate the resistance when the internal carotid is blocked with only a .25 cm radius.

2 | What happens to friction and resistance as you increase the length of the blood vessel?

3 | Explain the relationship between the radius of the blood vessel and resistance.

4 | Describe the relationship between viscosity, resistance, and flow.

5 | Comparing viscosity, blood vessel radius, and blood vessel length, which one do you think has the greatest impact on resistance? Explain your reason.

Materials Needed to Complete the Lab

- 30-ml (cc) syringes
- Rubber tubing of varying lengths—6", 10", 20"
- Rubber tubing of varying radii—1/8", 3/16", 1/4"
- Beakers
- Hose nozzle/connector
- Shampoo
- Conditioner
- Calculator
- Stopwatch

Lab Activity 1 | Exploring the Effects of Blood Vessel Length on Blood Flow, Pressure, and Resistance

As we grow, our blood vessels lengthen. As the vessel length increases, the resistance increases because of the greater amount of contact between the blood and the blood vessel lining. Therefore, friction increases as the length of the vessel increases. Once we reach adulthood, our vessel length stays consistent, except in the cases of weight loss and weight gain. With an increase in weight, blood vessels increase in length; with weight loss, vessel length decreases.

Activity Objectives

1 | To demonstrate and describe the effects of vessel length on flow rate and resistance.

2 | To calculate the relationship between pressure and flow.

3 | To describe conditions that can lead to blood vessel changes in the body.

In this activity, the different length tubes represent the different length blood vessels you would find in the body.

With your understanding of resistance and flow, create a hypothesis of what you believe will happen between the three different length tubes.

Hypothesis:

Materials Needed for Activity 1

- Three 30-ml (cc) syringes each filled with water
- Three lengths of rubber tubing—6", 10", 20"
- Hose nozzle/connector
- Calculator
- Stopwatch
- Beaker

Procedures

1 | Fill each of the syringes with 30 ml of water.

2 | Attach the rubber tubing to each syringe; make sure the tubing is over the syringe tip tightly to prevent it from sliding off.

3 | Start with the 6" tube.

4 | Place the free end of the tube in the beaker, making sure the tubing is not resting on the bottom of the beaker.

5 | One partner will control the stopwatch start and stop; the other partner will begin to push the water through the tubing attached to the syringe. Make sure to start the stopwatch and push the water through at the same time.

Please Note: The same person should control the syringe with each tube to ensure that there is consistency in the degree of force used to eject the fluid.

6 | When all the water has made it through the tube, stop the stopwatch and record the time in Table 23.1.

7 | Repeat Steps 1–6 above for each length of tube.

Table 23.1 | Time for Water to Empty into Beaker

Tube Length	Time for Water to Fill 30 ml
6"	
10"	
20"	

Observations

1 | Which length of tubing took the longest amount of time to empty the water into the beaker?

2 | Which length of tubing required more pressure to empty the water into the beaker?

Stop and Think Questions

1 | Did the measurements you observed match your hypothesis? If not, explain your new understanding.

2 | How would you relate the time differences to resistance and length?

3 | Based upon your answer in Question 2, create an explanation about the relationship between obesity, blood vessel length, and resistance.

4 | The relationship between blood flow and resistance is calculated using the following calculation.

$\boldsymbol{F = (P_A - P_V)/R}$

- Where F = Flow
- P_A = Arteriole Pressure
- P_V = Venous Pressure
- R = Resistance measured as mm Hg/ml/min
- Use this equation to answer the following question.

Renal blood flow is measured by placing a flow meter on an individual's left renal artery. Simultaneously, pressure probes are inserted in the left renal artery and left renal vein to measure pressure. Renal artery resistance is 50 mm Hg/ml/min. The pressure probes measure renal arterial pressure as 100 mm Hg and renal venous pressure as 10 mm Hg. What is the flow rate for the kidney's renal artery?

5 | Using the same numbers from the kidney question, increase the resistance by 10 and calculate what the new flow would be. Did it increase or decrease?

6 | What happens to flow when you increase the arterial pressure to 120 mm Hg and the venous pressure is 20 mm Hg? Use the *R* value provided in the original scenario.

Lab Activity 2 | Exploring How Blood Vessel Radius Affects Blood Flow and Resistance

The main method for controlling blood flow is through changing the blood vessel radius. To do this, the smooth muscle tissue found in the tunica media is either contracted to cause vasoconstriction or relaxed to cause vasodilation. When understanding the relationship between the radius and blood flow, consider the flow like lanes of highway traffic. On a highway there are slow lanes and fast lanes. The blood that is closest to the lining of the blood vessel will have increased contact with the vessel lining, generating friction and increasing resistance. This is the slow lane. Conversely, the blood in the center of the lumen will be less restricted because it is less likely to meet the vessel lining. This is the fast lane. This free moving flow of the blood at the center of the lumen is called **laminar flow.**

Try applying this concept with two different vessels. Let's say that vessel one is a large radius vessel or vasodilated and that vessel two is a smaller radius vessel.

1 | Which one is going to have the greater laminar flow?

2 | Which one is going to have greater resistance?

3 | In the space provided below, draw what you think each vessel's flow pattern would look like using narrow arrows for restricted flow and broad and bold arrows for laminar flow.

Activity Objectives

1 | To identify, calculate, demonstrate, explain, and/or describe how blood vessel radius affects flow rate.

2 | To explain how blood vessel radius affects resistance.

3 | To calculate the flow rate for varying tube radii.

In this activity tubes of varying radii represent vessels of different radii you would find in the body.

With your new understanding about the vessel radius, create your hypothesis of what will occur between the three tubes of different radii to resistance and flow.

Hypothesis:

Materials Needed for Activity 2

- Three 30-ml syringes filled with water
- Rubber tubing of different radii—1/8", 3/16", 1/4"
- Beaker
- Nozzle connector
- Stopwatch
- Calculator

Procedures

1 | Fill each of the syringes with 30 ml of water.

2 | Attach the rubber tubing to each syringe; make sure the tubing is over the syringe tip tightly to prevent it from sliding off.

3 | Start with the 1/4" radius tube.

4 | Place the free end of the tube into the beaker, making sure the tubing is not resting on the bottom of the beaker.

5 | One partner will control the stopwatch start and stop; the other partner will begin to push the water through the tubing attached to the syringe. Make sure to start the stopwatch and push the water through at the same time.

Please Note: The same person should control the syringe with each tube to ensure that there is consistency in the degree of force used to eject the fluid.

6 | When all the water has made it through the tube, stop the stopwatch and record the time in Table 23.2.

7 | Repeat Steps 1–6 above for the 3/16" and 1/8" radius tubing.

Table 23.2 | Time for Water to Empty into Beaker

Tube Radius	Time for water to fill 30 ml
1/4"	
3/16"	
1/8"	

Observations

1 | Which length of tubing took the longest amount of time to empty the water into the beaker?

____

2 | Which length of tubing required more pressure to empty the water into the beaker?

____

Stop and Think Questions

1 | Did the measurements you observed match your hypothesis? If not, explain your new understanding.

2 | How would you relate the time differences to resistance and radius?

When discussing blood flow, the term velocity is often used. **Velocity** is how quickly an object moves from one place to another and is often expressed in the units of cm/sec. In contrast, flow is the volume of a liquid or gas that is moving per unit of time. For blood flowing in a large vessel, flow is often expressed in the units of ml/min. The flow of blood in a vessel is related to velocity by the following equation.

$\boldsymbol{F = V \times A}$

- Where F = Flow
- V = Mean velocity
- A = Cross-sectional area of the vessel

$\boldsymbol{A = \pi r^2}$

- Where A = Cross-sectional area of a vessel
- r = Radius
- $\pi = 3.14$

3 | Calculate the flow rate in the ascending aorta with a radius of 1.2 cm and a mean velocity of 11 cm/sec.

4 | Using the same velocity given in Question 3, calculate the flow rate for each tube used in this activity. Record the flow in Table 23.3.

Table 23.3 | Calculated Flow Rate for Each Tube

Tube Radius Conversion Inches to Centimeters: 1" = 2.54 cm	Flow Rate
1/4" = 0.635 cm	
3/16" = 0.476 cm	
1/8" = 0.317 cm	

Lab Activity 3 | Exploring the Effects of Blood Viscosity on Blood Flow and Resistance

Viscosity is consistent in a healthy body. However, there are instances where blood's viscosity can be affected by a change in homeostasis. For example, the blood's viscosity can be affected by dehydration, or by an increase or decrease in erythrocyte levels. An increase in erythrocyte count (polycythemia) increases viscosity. Conversely, a decrease in erythrocytes (anemia) can decrease viscosity.

Objectives

1 | To describe how blood viscosity affects blood flow rate.

2 | To list the components that contribute to the viscosity of blood.

3 | To describe conditions that might lead to a change in blood viscosity.

In this activity you will use three fluids of different viscosities to examine the effects of viscosity on flow rate.

With your new understanding of viscosity, create your hypothesis of what will occur between the three fluid viscosities to resistance and flow.

Hypothesis:

Materials Needed for Activity 3

- Two 30-ml syringes
- Water
- Shampoo
- Conditioner
- Rubber tubing of different radii—1/4" and 1/8"
- Beaker
- Nozzle connector
- Stopwatch

Procedures

1 | Measuring the effect of viscosity with the 1/4" radius tubing.

a | Fill one of the 30-ml syringes with 30 ml of water.

b | Attach the 1/4" rubber tubing to this syringe; make sure it is over the syringe tip tightly to prevent it from sliding off.

c | Place the free end of the tube into the beaker, making sure the tubing is not resting on the bottom of the beaker.

d | One partner will control the stopwatch start and stop; the other partner will begin to push the water through the tubing attached to the syringe. Make sure to start the stopwatch and push the water through at the same time.

Please Note: The same person should control the syringe with each tube to ensure that there is consistency in the degree of force used to eject the fluid.

e | When all the water has made it through the tube, stop the stopwatch and record the time in Table 23.4.

f | Repeat Steps a–d using shampoo in the syringe rather than water.

g | Empty the shampoo back into the shampoo bottle.

h | Rinse the tube.

i | Repeat Steps a–d using conditioner in the syringe.

j | Empty the conditioner back into the conditioner bottle.

k | Rinse the tube.

2 | Repeat all of the steps above using 1/8" radius tubing.

Table 23.4 | Measuring the Effect of Viscosity with Tubing

Fluid	Time for 1/4" Radius Tubing	Time for 1/8" Radius Tubing
Water		
Shampoo		
Conditioner		

Observations

1 | Which fluid took the longest amount of time to empty into the beaker? ______________________

2 | Which diameter of tubing took the longest amount of time to empty into the beaker? ____________

Stop and Think Questions

1 | Did the measurements you observed match your hypothesis? If not, explain your new understanding.

2 | Thrombocytopenia is a decrease in platelets. What impact do you think this would have on viscosity?

3 | Explain the relationship between viscosity and flow.

4 | Consider the radius of the tube and viscosity, and then assess which factor you think had the greatest impact on the flow rate? Why?

5 | Predict what would occur to viscosity and flow during dehydration.

Laboratory 23 Review Questions

Hemodynamics

Name: ____________________

Putting it All Together

Critical Thinking Predictions

1 | Recall the heart is a pump, and we can relate the chambers' activity to specific phases of the cardiac cycle. Those phases are systole and diastole. When the chambers are **not** contracting/pumping, they are relaxed and filling with blood **(diastole).** When the chambers are contracting/pumping, they are ejecting blood into another chamber or artery **(systole).** Remember that cardiac output is calculated by heart rate (HR) × stroke volume (SV). Heart rate gives us a unit of time, beats per minute (BPM), and stroke volume gives us the amount of blood in ml/beat; therefore, cardiac output is measured as ml/min or L/min (there are 1000 ml in 1 L). Any increase or decrease in either stroke volume or heart rate affects the cardiac output.

Blood flow directly affects the stroke volume. Anything that affects blood flow, like resistance, affects stroke volume.

Predict what would happen to stroke volume and blood flow if there was an increase in resistance.

2 | Recall from Activity 1 we used the equation for flow, $F = (P_A - P_V)/R$, where F = Flow; P_A = Arteriole Pressure, P_V = Venous Pressure, and R = Resistance. We can simplify the $(P_A - P_V)$ to change in pressure (ΔP). Therefore, we can solve for $\Delta P = F \times R$ (Change in Pressure = Flow × Resistance).

We can develop a mathematical equation substituting cardiac output for flow to get $\Delta P = HR \times SV \times R$ which means heart rate, stroke volume, and resistance all have an impact on the blood pressure in the body. To maintain a normal blood pressure, heart rate, stroke volume, and resistance can be altered.

Predict what would happen to blood pressure if resistance was increased due to vasoconstriction of the arterioles.

3 | **Afterload** is the amount of resistance the left ventricle must overcome to force open the aortic semilunar valve, ejecting the blood into the ascending aorta. The greater the afterload, the harder it is to open the aortic semilunar valve, and less blood is ejected into the ascending aorta. **Stroke volume** is the amount of blood ejected into the ascending aorta during left ventricular systole. Stroke volume can be calculated by subtracting the End Systolic Volume (ESV) from the End Diastolic Volume (EDV). SV = EDV – ESV. EDV is the amount of blood in the ventricle when the ventricle is 100% filled. ESV is the amount of blood left in the ventricle after systole (contraction) ends. The greater the afterload, the larger the ESV.

Resistance is generated from the body's arterioles pushing blood back towards the heart and is called **peripheral vascular resistance.** When there is an increase in peripheral vascular resistance, afterload is increased.

a | Predict what would happen to the ESV if peripheral vascular resistance was increased.

b | Predict what would happen to cardiac output if the ESV amount was increased.

4 | Coronary artery disease is a form of atherosclerosis, where plaques form in the coronary arteries of the heart. This causes a decrease in the radius of the arteries due to the accumulation of plaque within the lumen of the arteries. Recall the right and left coronary arteries come off the ascending aorta, just above the aortic semilunar valve.

a | Predict what will happen to the resistance in the coronary vessels that contain plaques.

b | What will happen to the resistance on the aortic semilunar valve? How will this impact afterload? Explain.

c | How will this impact the ability of the heart to contract (with greater or less force)?

5 | Calcium channel blockers like verapamil are used to help the smooth muscle of the arteries relax and is often a treatment for coronary artery disease.

Explain how verapamil will affect resistance in the coronary arteries. How will this affect blood pressure in the body as a whole? How will this impact the ability of the heart to contract (with greater or less force)?

Terms to Understand for the Practical

- Blood flow
- Friction
- Afterload
- Resistance
- Viscosity
- Blood pressure
- Pressure gradient
- Poiseuille's law
- Vasoconstriction
- Vasodilation
- Laminar flow
- Velocity
- Albumin
- End systolic volume
- Peripheral vascular resistance
- Diastole
- Systole
- End diastolic volume
- Stroke volume

Formulas to Understand and Calculate for Practical

- $R = \frac{8\eta l}{\pi r^4}$
- $F = (P_A - P_V)/R$ **or** $\Delta P = F \times R$
- $F = V \times A$

24 Systemic Blood Vessels

The cardiovascular system includes a series of connected blood vessels known as ***arteries*** and ***veins.***

The arteries carry blood ***away*** from the heart to the various organs and tissues.

The veins drain blood from the organs and tissues and deliver it ***to*** the heart.

Arteries and veins both have three layers, or ***tunics,*** which compose their walls. Although they have the same layers, the walls of the arteries are much thicker, more muscular, and more elastic. This is due, in part, to the fact that arteries are under high pressure and must be able to expand, recoil and push the blood onto the next section of vessels. On the other hand, veins are under less pressure, are easily distensible, and can widen to pool large quantities of blood.

When examining blood vessels in the human cadaver, you should be able to notice the difference in the thickness and texture of the arteries versus veins.

Lab Activity

Using your lab manual, textbook, models, and other references, locate the following vessels on the human cadaver.

Lab Note: The number of vessels you will be able to locate will vary dependent upon the availability of the vessels for viewing on the human cadaver. Your instructor will notify you as to which vessels you will be responsible for learning.

Lab Tip: Often, the same vessel will be given a different name based upon the area of the body through which it passes. An example is that the subclavian artery (in the thoracic area) becomes the axillary artery as it passes into the armpit (axilla) and then it is named the brachial artery as it continues into the arm.

Arteries

A | Divisions of the Aorta

The aorta originates from the left ventricle of the heart. It is the largest artery in the body and is divided into several portions based upon its location. Find the following sections of the aorta.

Table 24.1 | Divisions of the Aorta

Vessel Name	Location
Ascending aorta	Initial portion of the aorta as it first ascends from the heart.
Aortic arch	Superior to the ascending aorta; forms a bend (arch).
Thoracic aorta	The portion of the aorta which is contained within the thoracic (chest) cavity.
Abdominal aorta	The portion of the aorta which is contained within the abdominal cavity. The aorta becomes the abdominal aorta immediately inferior to the diaphragm.

Lab Note: Both the thoracic aorta and abdominal aorta are included as part of the ***descending aorta*** which is the portion of the aorta after the aortic arch.

B | Head and Neck

There are three branches extending from the aortic arch which give rise to the arteries of the head and neck.

Table 24.2 | Head and Neck

Vessel Name	Location
Brachiocephalic artery (trunk)	A short artery located ***only* on the right side.** This is the first of the three branches extending from the aortic arch. It almost immediately gives rise to the right common carotid and right subclavian arteries.
Right common carotid	Branches medially from the brachiocephalic artery. Extends superiorly towards the head. The carotid arteries (right and left) are often used to check the pulse in the neck region.
Right subclavian	The more lateral branch of the brachiocephalic artery, it travels towards the right arm.
Left common carotid	This is the second (middle) branch off the aortic arch. It will supply the head.
Left subclavian	The third branch from the aortic arch, on the left side. Branching off the subclavian arteries (on both the right and left sides) are the small ***internal mammary*** (or ***internal thoracic***) ***arteries*** which can be used for coronary artery bypass surgery.
Vertebral	These are paired arteries which branch off both subclavian arteries and extend superiorly through the transverse foramina of the cervical vertebrae (and then through the foramen magnum) to supply the brain. Once in the brain, they unite to form the ***basilar*** artery.
Internal carotid	Paired arteries which branch from the common carotid arteries. These are the more posterior of the branches of the common carotid arteries. They tend to provide blood to areas *internal* to the skull such as the brain, eyes, and sides of the head and nose. The internal carotid arteries and branches of the basilar artery form the ***circle of Willis*** (or ***cerebral arterial circle***) which is a major circulatory system of the brain.
External carotid	These paired arteries also branch from the common carotid arteries. They supply much of the neck and external head (i.e., structures that are *external* to the skull).

C | Upper Limb

These are some of the major arteries which supply the arm.

Table 24.3 | Upper Limb

Vessel Name	Location
Axillary	Paired arteries which are extensions of the subclavian arteries at the armpit (axilla) area.
Brachial	Continuation of the axillary arteries in the upper arm. These vessels branch at the elbow into the ***radial*** and ***ulnar*** arteries.
Radial	The lateral branch of the brachial artery (on the thumb side). Near the distal end of the radial artery (at the wrist), it is covered with only fascia and skin; this makes a convenient location for measuring the radial pulse.
Ulnar	The more medial branch of the brachial artery (on the "pinkie" side).

D | Branches of the Abdominal Aorta

The abdominal aorta begins just inferior to the diaphragm and extends through the abdomen to a "V"-shaped branching of the common iliac arteries. The vessels below are given in order from superior to inferior as the abdominal aorta travels through the abdomen. (**Note:** Not all vessels branching from the abdominal aorta are listed.)

Table 24.4 | Branches of the Abdominal Aorta

Vessel Name	Location
Celiac trunk	A large, unpaired short vessel which divides into three branches supplying the stomach, spleen, pancreas, and liver. This is the first vessel encountered immediately as the abdominal aorta passes through the diaphragm.
Superior mesenteric	Another unpaired vessel which supplies the small intestine and part of the large intestine.
Renal	Paired arteries that feed the kidneys.
Inferior mesenteric	Single vessel off the aorta which supplies the large intestine (distal regions).
Common iliac	Branches formed at the inferior end of the abdominal aorta.

E | Pelvis and Lower Limb (Leg)

The common iliac arteries branch into the internal and external iliac arteries. The distal ends of the external iliac arteries then pass through the abdominal wall and enter the thigh where they are called the femoral arteries.

Table 24.5 | Pelvis and Lower Limb (Leg)

Vessel Name	Location
External iliac	Paired arteries that are extensions of the common iliac arteries. The more lateral of the two branches from the common iliac arteries.
Internal iliac	Paired arteries that branch from the common iliac arteries. These are more medial than the external iliac arteries and proceed deep into the pelvis. They supply the bladder and many pelvic structures.
Femoral	Once the external iliac arteries pierce the abdominal wall and enter the thigh, they become the femoral arteries. They supply many of the muscles of the thigh area.
Popliteal	At the posterior knee joint, the femoral arteries become the popliteal arteries. These arteries form branches that supply the lower leg (including ***posterior*** and ***anterior tibial arteries***).

Veins

A | The Venae Cavae

The two major veins delivering deoxygenated blood from the organs back to the right side of the heart are the superior and inferior venae cavae (vena cava = singular). The superior vena cava drains the upper portion of the body while the inferior vena cava drains the lower (inferior) portion of the body.

Table 24.6 | The Venae Cavae

Vessel Name	Location
Superior vena cava	Look for this short vessel (about 3") above the heart, near the aorta. It will be collapsed because of the thin walls of veins. An unpaired ***azygos*** vein is located on the posterior side of the superior vena cava. Look for it on the right side of the superior vena cava. There are several veins that empty into the azygos vein and which drain the thoracic and abdominal walls. It is suspected that the azygos vein can serve as a bypass vessel if the inferior vena cava becomes obstructed.
Inferior vena cava	This vessel can be located in the thoracic cavity if you lift up the heart. Look for it near the upper edge of the diaphragm and near the apex of the heart. Follow it superiorly as it passes into the heart. In addition, the inferior vena cava can be located immediately inferior to the diaphragm in the abdominal cavity. It is a large, collapsed tube extending parallel to the abdominal aorta. Note the difference between the structure of the abdominal aorta (an artery) and the inferior vena cava (a vein)!

B | Head and Neck

The veins of the head and neck all drain into the superior vena cava.

Table 24.7 | Head and Neck

Vessel Name	Location
Brachiocephalic	Paired veins that form V-shaped branches into the superior vena cava. Unlike in the arterial system, there are two brachiocephalic veins (but only one brachiocephalic artery).
Subclavian	These are lateral branches of the brachiocephalic veins. The subclavian veins drain blood from the arms, neck, and thoracic wall.
Internal jugular	Drain blood from the face, neck, and brain and empty into the subclavian veins. The internal jugular veins are located laterally to the internal and common carotid arteries. These are larger and more medial than the external jugular veins.
External jugular	These are paired veins that travel along either side of the neck and drain the facial muscles and scalp. They are smaller and more lateral than the internal jugular veins. The external jugular veins also empty into the subclavian veins. These veins are more superficial and may become visible when coughing or when a person is angry.

C | Upper Limb

The table below includes some of the major veins that drain primarily the upper arm. There can be anomalies (differences) from person to person in some of the veins of the arm. Veins are numerous and form many tributaries as well as anastomoses (connections to each other). The arm veins have valves built in to ensure the flow of blood upward towards the heart.

Notice that many of the veins of the arm correspond to the arteries of the upper arm. Look for them in the same locations and remember that the veins should be more collapsed.

Table 24.8 | Upper Limb

Vessel Name	Location
Axillary	Paired veins which empty into the subclavian arteries at the armpit (axilla) area.
Brachial	Upper arm vein that continues into the axillary vein. The ***radial*** and ***ulnar veins*** of the forearm empty into the brachial veins. The brachial, radial, and ulnar veins are all deep veins of the arm.
Cephalic	The cephalic veins are lateral, superficial, paired veins that empty blood from the upper arm into the axillary veins.
Basilic	The basilic veins are medial, superficial, paired veins that drain the upper limbs. The basilica and brachial veins merge in the axillary area to form the axillary veins. Anterior to the elbow, the cephalic and basilic veins are connected by the ***median cubital vein.*** This is the vein that is preferred for venipuncture (blood drawing), transfusions, or injections.

D | Branches of the Inferior Vena Cava

The veins listed below are located in the abdominal cavity and they empty their blood into the inferior vena cava. They are listed from superior to inferior. (**Note:** Not all veins in the abdominal cavity are listed.)

Table 24.9 | Branches of the Inferior Vena Cava

Vessel Name	Location
Hepatic	Located at the superior aspect of the abdominal portion of the inferior vena cava (look for them near to where the inferior vena cava pierces the diaphragm). These veins deliver blood from the liver into the inferior vena cava.
Renal	Paired veins that drain the kidneys.
Common iliac	Branches at the distal end of the inferior vena cava. They join to create the inferior vena cava.

E | Pelvis and Lower Limb (Leg)

The common iliac veins branch into the internal and external iliac veins. The distal ends of the external iliac veins pass through the abdominal wall and enter the thigh where they are called the femoral veins.

The veins of the leg also have many valves to ensure blood flow back to the heart.

Table 24.10 | Pelvis and Lower Limb (Leg)

Vessel Name	Location
External iliac	Paired veins that empty into the common iliac veins. These are lateral.
Internal iliac	Paired veins that also empty into the common iliac veins. These are more medial than the external iliac veins and proceed deep into the pelvis. They empty blood from the bladder and many pelvic structures.
Femoral	These veins are located in the thigh and empty into the external iliac veins. They are extensions of the external iliac veins. They drain much of the leg.
Great saphenous	This is the longest vein in the body. It originates at the foot and ascends to empty into the femoral vein at the groin. Look for it on the medial side of the leg. Because it is a superficial vein, it may be torn or broken during dissection or handling in the lab. The great saphenous veins are often surgically removed and used as grafts for coronary bypass surgeries. This vein has numerous valves and is prone to varicosities if the valves become incompetent (weak). The great saphenous vein is also a common site for the administration of intravenous fluids when other veins have collapsed.

F | Hepatic Portal System

The inferior vena cava does not receive blood from veins that arise in the pancreas, spleen, gastrointestinal tract (stomach and intestines), or the gall bladder. Instead, the blood from these veins, which contains absorbed nutrients from digestive processes, empties into a special vein, the ***hepatic portal vein;*** the hepatic-portal vein then transports the blood to the liver. Once in the liver, the blood is processed. In the liver, blood nutrients may be extracted for use, stored, or eliminated from the body; toxic substances may be detoxified or stored, and bacteria may be destroyed. Next, the newly-processed blood in the liver enters the ***hepatic veins*** which then empty the blood into the inferior vena cava.

Table 24.11 | Hepatic Portal System

Vessel Name	Location
Superior mesenteric	This vein drains the pancreas, the small and large intestines, and parts of the stomach. The blood is delivered to the hepatic portal vein. This vein and the splenic vein merge to form the hepatic portal vein.
Splenic	This major vein empties into the hepatic portal vein and drains the spleen, stomach, and part of the pancreas. It merges with the superior mesenteric vein to form the hepatic portal vein.
Inferior mesenteric	This vein drains a portion of the large intestine and merges with the splenic vein, near the entry of the splenic vein into the hepatic portal vein.
Hepatic portal vein	The hepatic portal vein empties into the liver.

Required Blood Vessel List for Practical

After completion of the blood vessel labs, you should be able to identify and correctly spell the following arteries and veins on the human cadaver *if they are viewable* on the particular cadaver being utilized. Your instructor will inform you of any changes in the list (possible additions or deletions).

Table 24.12 | Required Blood Vessel List for Practical

	Arteries	Veins
Thorax	Ascending aorta	Superior vena cava
	Aortic arch	Brachiocephalic (right and left)
	Descending aorta	Internal jugular
	Brachiocephalic trunk	External jugular
	Common carotid (right and left)	Subclavian (right and left)
	Subclavian (right and left)	
Arm	Axillary	Axillary
	Brachial	Brachial
Abdomen	Abdominal aorta	Inferior vena cava
	Renal	Renal
Hip/Leg	Common iliac	Common iliac
	External iliac	External iliac
	Femoral	Femoral
	Deep femoral	Deep femoral
		Great saphenous
Chest Wall	Internal thoracic (mammary)	Internal thoracic (mammary)

Arteries

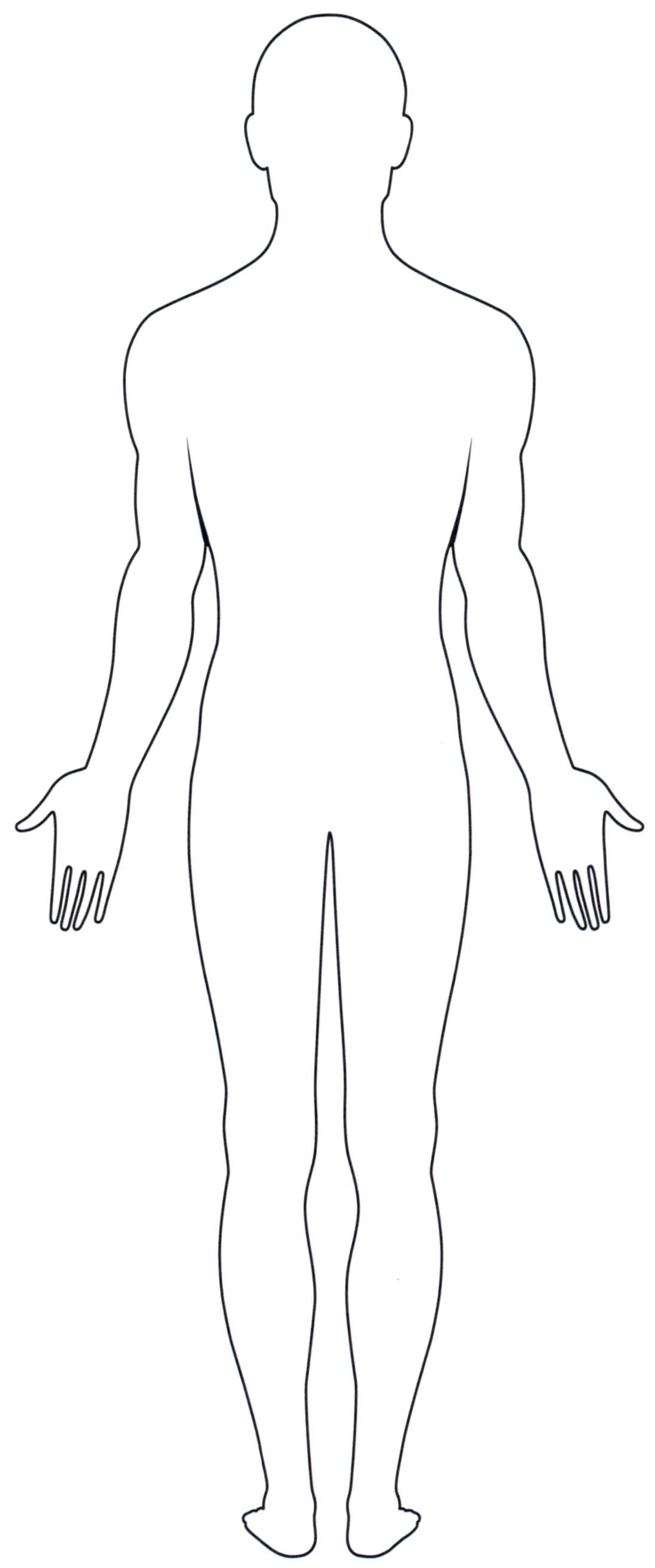

Veins

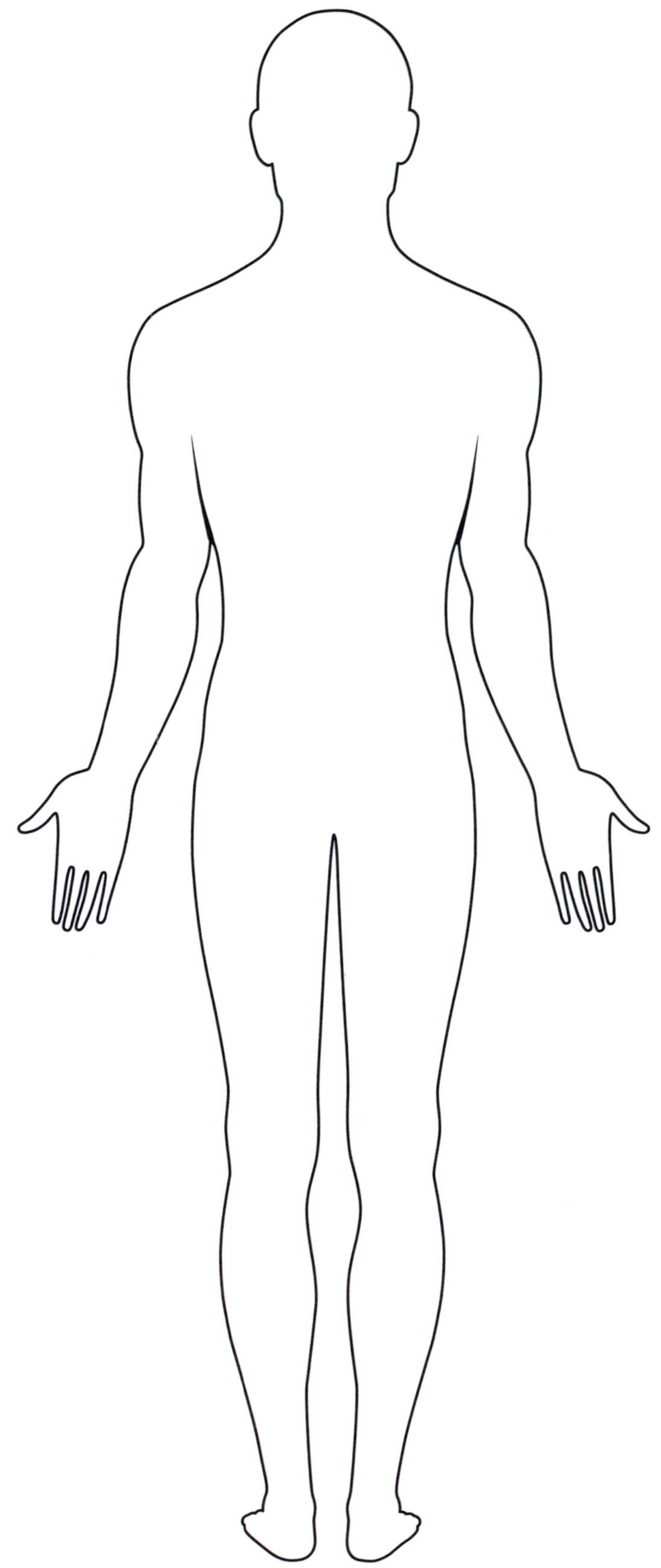

Laboratory 24 Review Questions

Systemic Blood Vessels

Name: ______________________________

1 | List the three major arteries arising from the aortic arch.

a |

b |

c |

2 | The brachiocephalic trunk is found on the __________.

a | right side   b | left side   c | both sides

3 | Which of the following is **not** a section of the aorta?

a | arch   c | iliac   e | abdominal

b | descending   d | ascending

4 | Place the following arteries in order, beginning with the artery closest to the heart and ending with the artery farthest from the heart.

________ 1. (closest)	a | brachial
________ 2.	b | right subclavian
________ 3.	c | aortic arch
________ 4.	d | axillary
________ 5.	e | brachiocephalic trunk
________ 6. (farthest)	f | ulnar

5 | Above the diaphragm, the aorta is known as the ________________ aorta; below the diaphragm, it is called the ________________ aorta.

6 | The great saphenous vein drains into the ________________ vein which empties into the ________________ vein and then empties into the inferior vena cava.

7 | The jugular veins and the subclavian veins all empty into the ________________ veins.

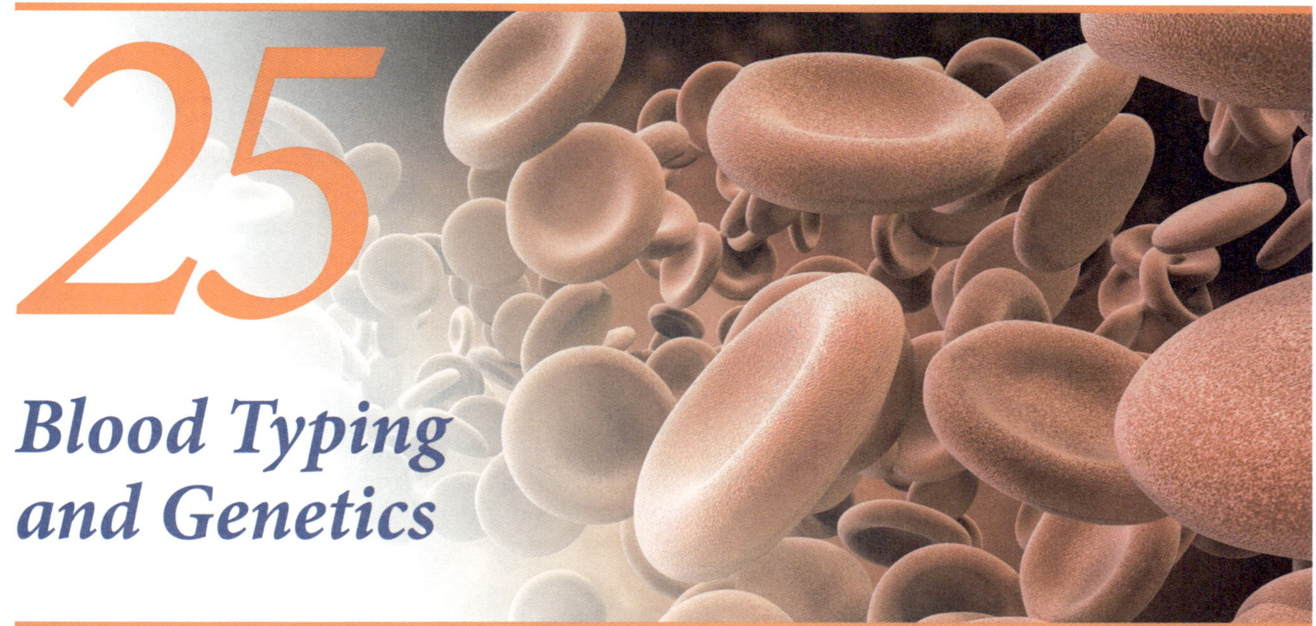

25 Blood Typing and Genetics

A | The ABO System

The most common four basic types of human blood are A, B, AB, and O. Blood type is based on the presence or absence of certain ***antigens*** (proteins) found on the surfaces of the ***erythrocytes*** (red blood cells). An antigen is a cell surface marker, which acts as a fingerprint enabling a cell to be one of a kind and also to be identified by other cells. Antigens are proteins found on all cells; however, the specific types of antigen found on the surface of the red blood cell is known as an ***agglutinogen*** because it has the ability to cause ***agglutination*** or clumping when reacting with specific antibodies (agglutinins). In the ABO typing system, the human red blood cell agglutinogens (RBC antigens) are A and B. The types of your red blood cell agglutinogens are inherited from your parents.

Antibodies are large proteins which interact with antigens and ultimately destroy them. There are many types of antibodies in the body, the most familiar being associated with the immune system. You may be aware of antibodies that are specifically formed in the body to fight various kinds of infections caused by the antigens on the surface of bacterial cells. Red blood cell antibodies are somewhat different, however, and are referred to specifically as ***agglutinins.***

In most instances, an individual develops antibodies after exposure to a foreign (non-self) antigen. But the ABO system is unique because the plasma has predetermined antibodies that an individual begins to develop shortly after birth (reaching adult levels around 8–10 years old). These plasma antibodies that are produced will react to the RBC antigen that the individual **does not** have because a person would not want to develop an antibody against his/her own antigens. In other words, if an individual has the A antigen, then that individual will develop anti-B antibodies (agglutinins). If the individual is born with B antigens on the surface of his/her red blood cells, then he/she will develop anti-A antibodies (agglutinins). Some people are born with both A and B antigens on their red blood cells; in this case, they would not develop antibodies against either antigen. Still other people are born with neither the A or B red blood cell antigens; these people are considered to be blood type O, and they develop both anti-A and anti-B antibodies (agglutinins).

Table 25.1 enables you to see the antigens and antibodies of the four ABO blood types. Remember that you develop the antibody against the antigen that you **do not** have. When considering compatibility between blood donors and recipients (those who receive blood), you should keep track of two items: **1) the antigens located on the red blood cells of the donor,** and **2) the antibodies present in the plasma of the recipient.**

Table 25.1 | Antigens and Antibodies of the Four ABO Blood Types

Blood Type	Antigens (Agglutinogens) on Red Blood Cells	Antibodies (Agglutinins) in Plasma	Blood Types Who They Can Donate To	Blood Type(s) Who They Can Receive Blood From
A	A	Anti-B	A, AB	A, O
B	B	Anti-A	B, AB	B, O
AB	A and B	Neither anti-A nor anti-B	AB	A, B, AB, and O
O	Neither A nor B	Both anti-A and anti-B	A, B, AB, and O	O

If an incompatible blood type is given, the result could be red blood cell destruction, agglutination, and other difficulties. For example, if an individual with blood type B is given blood type A, the recipient's anti-A antibodies will cause agglutination of the type A antigens.

B | The Rh Typing System

Another common blood typing system is the Rh system. Each ABO type is considered to be Rh positive (Rh^+) or Rh negative (Rh^-). The Rh factor (also called the D factor) was first isolated from the Rhesus monkey species, thus the abbreviation Rh. The majority of people are Rh^+. If a person has the Rh antigen present on his/her red blood cells, then he/she is considered to be Rh^+. If a person does **not** have the Rh antigen, then he/she is considered to be Rh^-.

Similar to the ABO system, a person's plasma can contain the antibody against the antigen which he/she **does not** have. Unlike the ABO system, in the Rh blood system individuals are not born with predetermined antibodies that automatically develop after birth.

An Rh^+ individual will **not** have anti-Rh antibodies because the antibodies would attack his/her Rh antigens. However, there are situations where an Rh^- person (one who does not have the Rh antigen) can develop the anti-Rh antibody. Normally, the anti-Rh antibody is not present in the plasma of any Rh^- individual. But when exposed to the Rh^+ antigen, the Rh^- person will develop the anti-Rh. For example, if an Rh^+ blood transfusion were given to an Rh- person, then that Rh- person would develop anti-Rh antibodies because that person's immune system will recognize the "foreign" Rh antigen. (This is similar to what occurs with a vaccination. For example, people aren't born with antibodies for polio but will develop the antibodies after receiving vaccinations that expose them to the polio antigen.) The Rh^- recipient is now ***sensitized*** to Rh positive blood (i.e., he/she now has anti-Rh antibodies in his/her plasma). Subsequent Rh^+ blood transfusions could cause the anti-Rh of the Rh^- person to agglutinate the Rh^+ blood, followed by ***hemolysis*** (bursting) of the red blood cells.

Please note that if an Rh^- person is sensitized to Rh^+ blood and develops the anti-Rh, the Rh^- person's blood type **does not** change to Rh^+. He/she is still Rh^- but now has the anti-Rh in the plasma. (Using the vaccination example above, if a person receives polio vaccines, his/her blood type doesn't change but he/she will have polio antibodies in his/her blood plasma.)

C | Rh Fetal/Maternal Incompatibility

A problem with Rh incompatibility may occur between a mother who is Rh^- and a fetus that is Rh^+. Under normal circumstances, fetal and maternal blood does not mix. However, there may be occasions when some of the fetal Rh^+ antigens can enter the mother's bloodstream. Once there, the mother's system will become sensitized to the presence of the foreign Rh antigens, and she will begin to develop a population of anti-Rh antibodies. This usually does not occur until the end of the pregnancy, so the antibody population does not increase enough to harm the fetal red blood cells. However, once labor and deliver occur, the resultant bleeding between uterus and placenta allows many fetal Rh^+ antigens to escape into the mother's circulation, and she builds up a high titer (amount) of anti-Rh antibodies. When antibodies are produced, they also develop a high degree of memory in case of future exposure.

This first Rh^+ fetus is not affected, but any subsequent Rh^+ fetuses could be adversely affected. In subsequent pregnancies, the mother's anti-Rh antibodies can cross the placenta and enter the fetal circulation. The interaction between the mother's anti-Rh antibodies and the fetal Rh^+ antigens causes agglutination and hemolysis (bursting and release of hemoglobin) of the fetal red blood cells. The fetus may be born severely anemic and may require transfusions. In some cases, the fetus can die as a result of the reaction. This condition is known as ***hemolytic disease of the newborn (or erythroblastosis fetalis).***

This condition may potentially occur only if the mother is Rh^- and the fetus Rh^+ and the mother has become sensitized. If the Rh^- mother is carrying an Rh^- fetus, there is no problem because those blood types are compatible. An Rh^- mother could be sensitized and build up anti-Rh antibodies from past pregnancies, through a blood transfusion, or from a past miscarriage or abortion of an Rh^+ fetus. Therefore, it is important to inform the physician if any of those situations have occurred. Incidentally, there is no incompatibility between an Rh^+ mother and an Rh^- fetus. Can you figure out why?

Recall that the mother does not produce the anti-Rh antibodies until the fetal Rh^+ antigens have sensitized her immune system. To prevent this from occurring, Rh^- mothers are given an injection of a product called Rho-Gam™, which is actually a population of anti-Rh antibodies. When these are given to the mother, they tie-up any of the few fetal Rh^+ antigens that may have leaked into her bloodstream, thus preventing her own immune system from being sensitized. She does not develop any of her own anti-Rh antibodies, and future Rh+ pregnancies are safe.

Lab Activity

ABO and Rh Blood Typing

As you know, you develop the antibody opposite to the red blood cell antigen that you have. If you are blood type A, then you have anti-B antibodies. However, when determining blood type in the laboratory setting, the technique employed is just the opposite. Suppose you want to see if an unknown drop of blood has the A antigen. What would you do? You would add a drop of anti-A antibodies to the droplet of blood. If A antigens were present on the red blood cells in the blood sample, then the anti-A antibodies would clump with the A antigens—proof that the blood is type A.

In this activity, you will determine the ABO and Rh blood types of four samples of unknown blood type using materials and procedures from *Ward's Simulated Blood Typing Kit™*.

Lab Note: This blood typing activity does **not** utilize real blood or blood sera. However, the procedures used are the same as those used to type actual human blood.

Materials Needed to Complete this Lab

- 4 plastic blood typing trays
- Toothpicks
- 4 vials of unknown blood samples:
  - Mr. Smith
  - Ms. Jones
  - Mr. Green
  - Ms. Brown
- Wax pencil for marking typing trays with blood sample names
- Anti-A antibody serum
- Anti-B antibody serum
- Anti-Rh antibody serum

Lab Note: For this activity, you will be working in groups of 3–4 people.

Procedure

1 | Using the wax pencil, label each of the 4 plastic blood typing trays as follows:

Mr. Smith – tray #1
Ms. Jones – tray #2
Mr. Green – tray #3
Ms. Brown – tray #4

2 | Place 2 drops of Mr. Smith's blood in the A, B, and Rh wells of tray #1.

3 | Place 2 drops of Ms. Jones's blood in the A, B, and Rh wells of tray #2.

4 | Place 2 drops of Mr. Green's blood in the A, B, and Rh wells of tray #3.

5 | Place 2 drops of Ms. Brown's blood in the A, B, and Rh wells of tray #4.

6 | Add 2 drops of anti-A antibody serum in the "A" wells of all four trays.

7 | Add 2 drops of anti-B antibody serum in the "B" wells of all four trays.

Lab Note: The anti-B antibody serum often takes longer to react than the other antibody sera. Be patient!

8 | Add 2 drops of anti-Rh antibody serum in the "Rh" wells of all four trays.

9 | Using **separate** toothpicks for each sample, stir the blood and antibody serum to thoroughly mix.

Lab Note: You **must** use separate toothpicks for each well on each sample in order to prevent cross-contamination of the samples!

10 | Any clumping or agglutination or cloudiness will indicate a positive reaction.

11 | Record your observations in Table 25.2.

Laboratory 25
Results of Blood Typing Tests

Name: ______________________________

Table 25.2 | Results of Blood Typing Tests

	Anti-A Serum	Anti-B Serum	Anti-Rh Serum	Blood Type	Observations
Tray #1 Mr. Smith					
Tray #2 Ms. Jones					
Tray #3 Mr. Green					
Tray #4 Ms. Brown					

Lab Note: In the boxes labeled "Anti-A Serum," "Anti-B Serum," and "Anti-Rh Serum" indicate if there was a **reaction** or **no reaction.** In the box "Observations" write your observations for each sample (e.g., cloudy, clumping, no change, etc.).

Questions Concerning ABO and Rh Typing Activity

Please note that Questions 1–4 below are referring to *both* the ABO and Rh systems! Answer the questions accordingly.

1 | Mr. Green has what kind of red blood cell antigens (agglutinogens)?

2 | What kind of ABO and Rh antibodies (agglutinins) does Mr. Green have in his plasma?

3 | Ms. Jones needs a blood transfusion. What kinds of blood can she safely receive?

4 | Ms. Brown wants to donate blood. What ABO/Rh blood types could receive her blood?

5 | What is the difference between an agglutinogen and an agglutinin?

6 | Why is it necessary to match the donor and recipient's blood before a blood transfusion?

7 | What happens to red blood cells that agglutinate?

8 | What determines the blood type to which each person belongs?

9 | How could an Rh^- person develop Rh agglutinins?

10 | Why can Rh^+ blood be given only once to a non-sensitized person who is Rh^-?

11 | Explain **specifically** what happens in hemolytic disease of the newborn (erythroblastosis fetalis). In what situations might it occur?

D | The Genetics of Blood Type

To understand the genetics of blood typing, it is necessary to understand some terms associated with genetics in general.

An ***allele*** is a gene that occupies a position on a chromosome. Alleles are inherited in pairs (one allele from each parent) with some alleles being ***dominant,*** some ***co-dominant,*** and some ***recessive.*** Dominant genes (usually indicated by upper-case letters) are always expressed over recessive genes (usually indicated by lower-case letters).

An example is hair length in cats. The allele for short hair is dominant and will be indicated by the letter "S." Long hair genes are recessive and indicated by the letter "s." An offspring of two parents can inherit several combinations of the two, depending on what genes the parents have. As an example, assume a kitten inherits the genetic combination "SS." This cat will have short hair. If a kitten inherited the genetic combination "Ss," it also would have short hair because the short-haired gene (S) dominates the recessive long-haired gene (s). In order for a cat to have long hair, both alleles would need to be the recessive "s" gene (ss = long hair).

When a pair of alleles is inherited, they may be the same or different as in the above examples. If a kitten had inherited the allele combination of SS, or ss, then the members of the pair would be said to be the same or ***homozygous.*** In the case of Ss, the genotype is ***heterozygous*** (the two genes are different) with a dominant short hair allele and a recessive long hair allele.

Any trait that is expressed (observable) is referred to as ***phenotype;*** in the example above, the phenotype is hair length in cats. The specific traits carried in the pair of alleles (what genes the individual has) are referred to as ***genotype.*** In the examples above, the genotypes are SS, Ss and ss. The genotype (genetic makeup) determines the phenotype (the actual observable trait).

Determination of genetic inheritance from parents to their offspring is typically done using a diagram called a ***Punnett square.*** Across the top of the square, the alleles of the mother (the genes her eggs would carry) are placed. Along the left side of the box, the alleles of the father (the genes his sperm would carry) are placed. By dropping each allele down and over into the corresponding boxes, the possible genotypes for all offspring are obtained.

To the right is a Punnett square using the hair length in cats as an example. For this example, let's assume we have a homozygous short-haired mother cat (genotype would be SS) and a heterozygous short-haired father cat (genotype would be Ss).

Father's Alleles \ Mother's Alleles	S	S
S	**SS**	**SS**
s	**Ss**	**Ss**

What kinds of hair could the kittens of these two parents have? The possible genetic combinations of the kittens are in the 4 center boxes. Looking at those boxes, you can see that the possible genotypes are SS, SS, Ss and Ss. In **all** cases, the genotypes would result in short-haired kittens because the S gene dominates.

The genes for blood types are inherited in the same way. The alleles for the A and B antigens are both dominant over the O allele. If **both** A and B alleles are present, neither one dominates the other; therefore, we refer to them as ***co-dominant,*** and they are both equally expressed if they occur together. In that case, the blood type would be AB. The allele for blood type O is recessive. Since an individual inherits an allele from both parents (and both parents have allele pairs that they also inherited from their parents), an individual may inherit different allele combinations (genotypes) for each blood type.

Table 25.3 | Possible Genotypes for Blood Types

Phenotype	Possible Genotypes
A	AA, Ao
B	BB, Bo
AB	AB
O	oo

As noted in the table above, there can be two possible genotypes for blood type A: AA and Ao, even though in the second case a recessive allele from blood type O is found. The same is true for blood type B. Type AB has only one possible combination, a heterozygous pair which are both expressed. Type O also has only one combination of alleles—a homozygous recessive pair (oo). Anytime a recessive trait is expressed, both alleles must be present.

Let's try a Punnett square using the ABO system. Assume the mother is heterozygous A (genotype Ao) and the father is heterozygous B (genotype Bo). Can you determine the possibilities of blood types for their children?

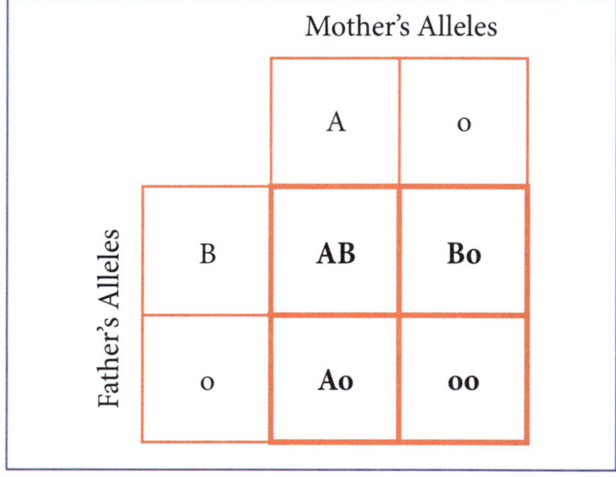

The A and o alleles at the top of the chart are contributed by the mother (these would be the genes for blood type in her eggs). The B and o alleles to the left of the chart are contributed by the father (these would be the genes for blood type in his sperm). By dropping down and over the alleles from the mother's eggs with the alleles from the father's sperm, the results are the possible blood types of their children—shown in the four boxes in the center of the Punnett square. Thus, if the parents are genotype Ao and Bo, the resulting possible blood types (phenotypes) for their children are A, B, AB, or O.

There are many alleles for the Rh factor. However, for simplification for this exercise, the allele for Rh^+ is dominant over the allele for Rh^-. To further simplify the genetics, we will use + for the Rh^+ allele and – to indicate an Rh^- allele. An Rh^+ individual could have two possible genotypes: ++ or +–; an Rh^- individual will have only one possible genotype – – since the Rh^- type is recessive. Work the Punnett squares for Rh factors as you would for the ABO typing.

Table 25.4 | Genotypes for Rh Factors

Phenotype	Possible Genotypes
Rh+	++ or +–
Rh^-	– –

Example of an Rh Punnett square:

Assume that the mother is heterozygous Rh^+ (genotype +–) and the father is Rh^- (genotype – –). What are the possibilities for Rh positive or Rh negative blood for their children?

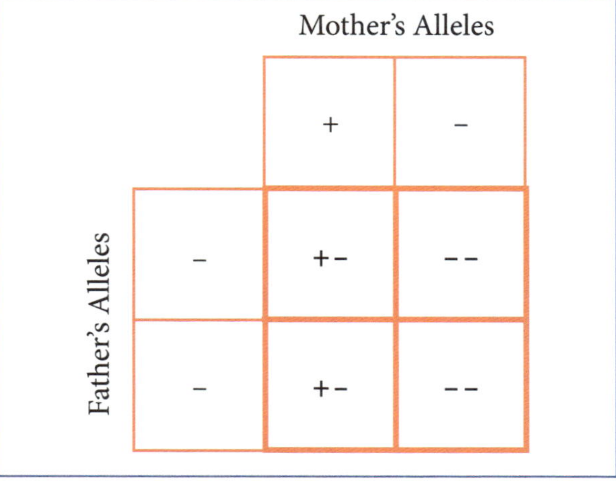

There is a 50% chance that their children could be Rh^+ (because of the two +– genotypes in the squares above), and a 50% chance their children could be Rh^- (because of the two – genotypes in the squares above).

Using the information above and working with Punnett squares, answer the questions on the following pages.

Laboratory 25 Questions Concerning the Genetics of Blood Type

Name: ____

ABO Blood Typing Problems

1 | What are the possible blood types of the children of a type B man and a type O woman? Is it possible for them to have type O children? **Explain** your answers (a simple "yes" or "no" answer is not sufficient). Use the Punnett squares below to help in your explanation.

2 | If a type A man and a type A woman have children together, what are the possible blood types of their children? Is it possible for them to have a type AB child? Is it possible for them to have a type O child? **Explain** your answers (a simple "yes" or "no" answer is not sufficient). Use the Punnett squares below to help in your explanation.

3 | Could a man with an AB blood type be the father of a type O child? **Explain** your answer (a simple "yes" or "no" answer is not sufficient). Use the Punnett square below to help in your explanation.

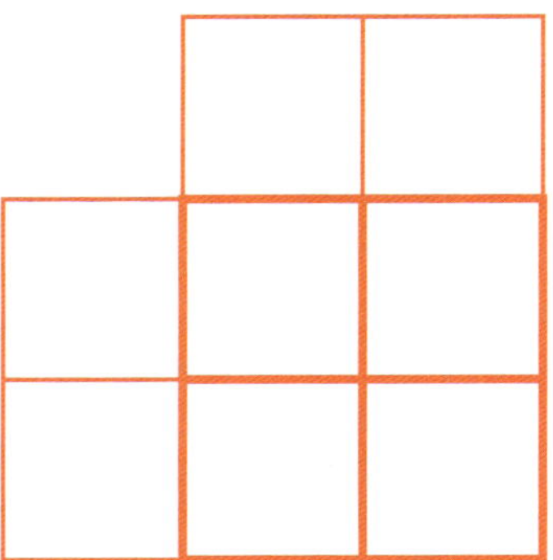

4 | What would be the **phenotypes** of the children who had a type A father and a type AB mother? What are the possible **genotypes** of the children? **Explain** your answers, using the Punnett Squares as an aid.

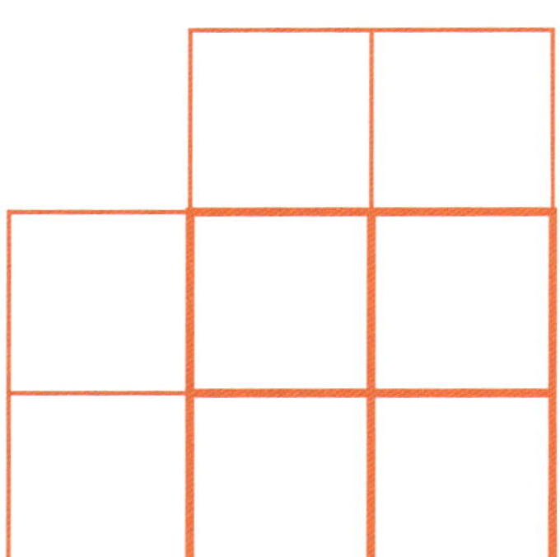

5 | What are the possible blood types of children produced by Ms. Jones and Mr. Green (these are two of the people whose blood type you determined in the blood lab)?

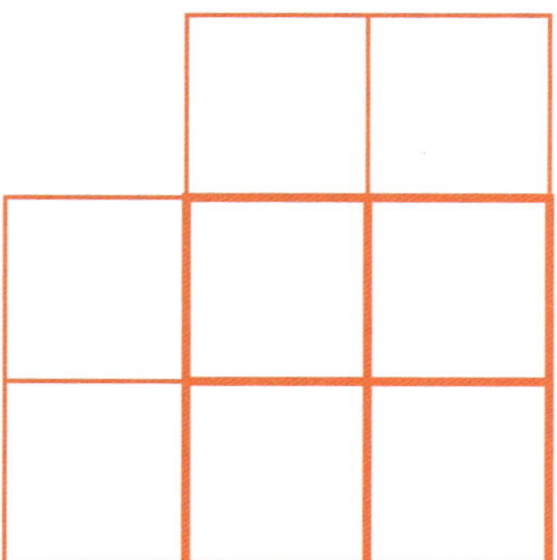

6 | What are the possible genetic combinations of offspring when the blood types of the parents are A and B? Use all possible genotypes for the parents. You will have to do four Punnett squares to answer this question. Put the percent probabilities next to each square.

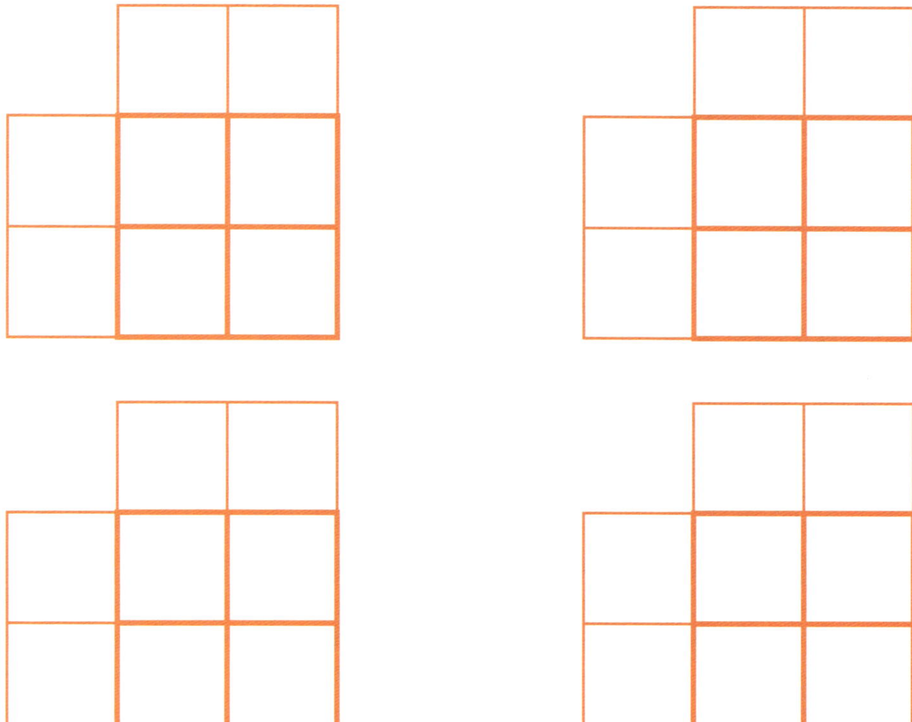

7 | Jane Doe has blood type O and her husband John Doe has blood type A. John's mother had blood type O. What are the possible blood types of their offspring? For this problem, you will need to determine how many Punnett squares are required to answer the question. You will need to draw your own Punnett square(s) to determine the answer.

Rh Blood Typing Problems

8 | If Mr. Smith and Ms. Brown produced offspring, what would the chances be for Rh^+ or Rh^-? (Do the Punnett squares for Rh only—not the ABO type.)

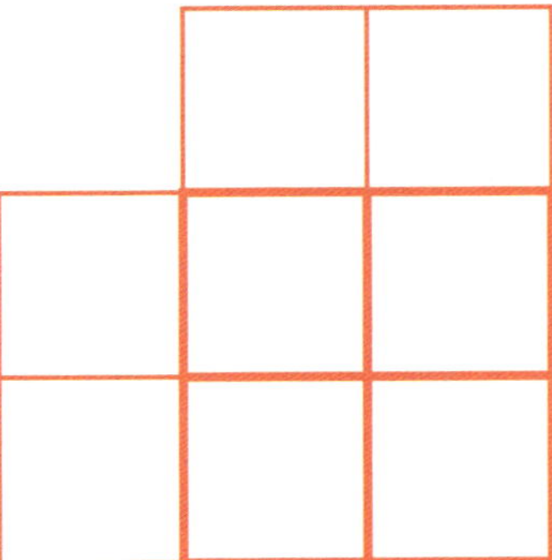

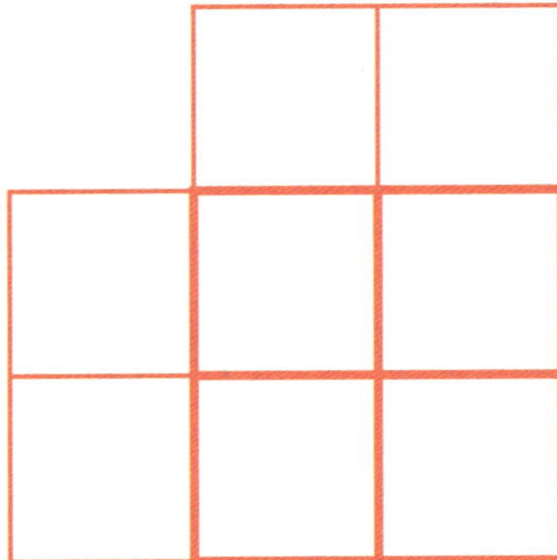

9 | What would be the possibilities for a Rh^+ or Rh^- children if the mother is homozygous Rh^+ and the father is homozygous Rh^-? For this problem, you will need to determine how many Punnett squares are required to answer the question. You will need to produce your own Punnett square(s) to determine the answer.

10 | Can two parents who are Rh^+ produce an Rh^- offspring? Work the Punnett squares and then use that information to explain your answer.

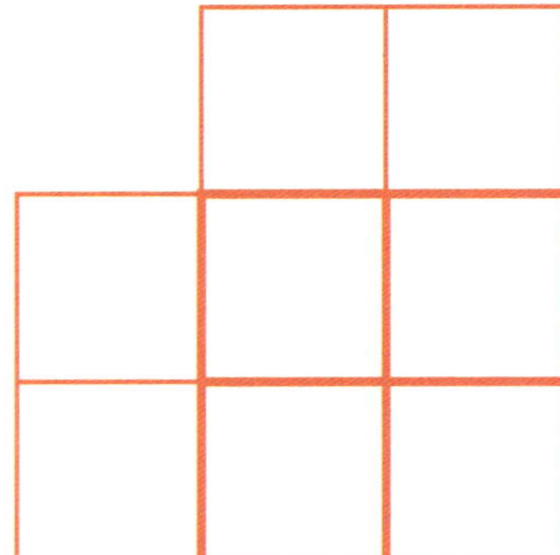

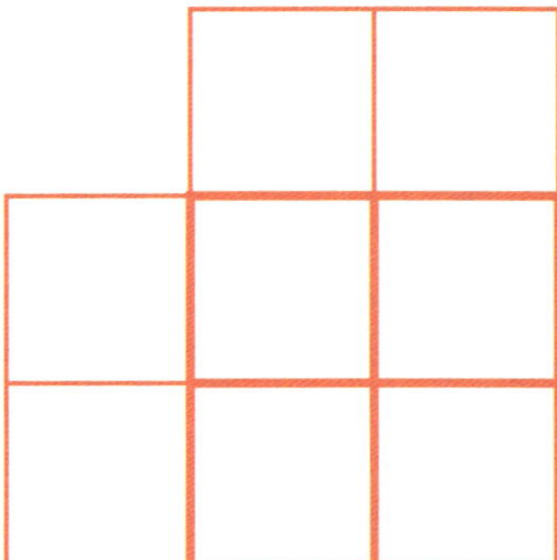

Terminology Review for Genetics of Blood Typing

After completion of the blood typing and genetics lab, you should understand the following terminology for your blood typing practical/quiz.

- ☐ **Chromosome:** Condensed and coiled DNA strands. Humans have 23 pairs of chromosomes (46 total).
- ☐ **Gene:** Sequence of nucleotides that hold codes for a specific protein. Located in specific locations on chromosomes.
- ☐ **Allele:** Alternate form of a gene found on each of the chromosomes in a pair.
- ☐ **Dominant allele/trait:** A gene that has a physical characteristic that usually is displayed/expressed when present in a person's DNA.
    - Use a capital letter to denote the dominant allele. Example: **B** = brown hair.
- ☐ **Recessive allele/trait:** A gene that is overpowered by the presence of a dominant trait.
    - Can be expressed if both alleles are recessive.
    - Use a lowercase letter to denote the recessive allele. Example: **b** = blond hair.
- ☐ **Homozygous:** Both alleles are the same, either dominant or recessive.
    - **BB** = homozygous dominant for brown hair
    - **bb** = homozygous recessive for blond hair
- ☐ **Heterozygous:** Alleles are different with one being dominant and one recessive.
    - **Bb** = heterozygous dominant for brown hair
- ☐ **Co-dominance:** Occurs when both alleles are dominant but for different characteristics; the result is both are expressed, creating a new characteristic. Example is AB blood type.
- ☐ **Genotype:** The alleles that are present in a person; the genetic recipe. Can be either homozygous (AA) or heterozygous (Aa).
- ☐ **Phenotype:** The physical expression of the genotype; the traits that you can observe. Examples: blue eyes, brown hair, blood type.

26

The Immune System

Part of the function of the immune system is to protect the body from foreign invaders, such as bacteria and viruses. The lymphocytes are essential components of this response. T lymphocytes are produced in the red bone marrow and must spend time in the thymus gland in order to become functional cells of the immune system. There are several types of T lymphocytes. They are involved in cell-mediated immunity and also play a role in antibody-mediated immune responses. B lymphocytes are produced and mature in the red bone marrow. They are involved in antibody-mediated immunity. Activation of B and T cells is a two-step process—meeting the foreign antigen, followed by confirmation.

A | Antibody-Mediated Immune Response

Antibody-mediated immune responses involve B lymphocytes. However, specific T cells, known as helper T cells are also required in antibody-mediated responses. The T cells will be introduced in the next section.

Antibody-mediated immune responses are initiated by the recognition of specific foreign antigens (such as bacteria or viruses) by ***inactive B lymphocytes.*** The B lymphocytes will attach to the antigens and begin to divide and produce ***active B cells*** (or ***plasma cells***) and ***memory B cells.*** The plasma cells secrete ***antibodies*** into the blood. These antibodies will bind to the antigen, forming the ***antigen-antibody complex,*** which will render the antigen inactive by a variety of mechanisms. ***Memory B cells*** are produced and will remain in the blood long term. They become activated the next time that particular antigen is encountered and will help to ensure that the subsequent response is swift and immediate.

B | Cellular-Mediated Immune Response

Cellular-mediated immune responses involve T cells. If a foreign antigen (such as bacteria or viruses) enters the body, phagocytic cells will engulf and partially digest the antigens. These phagocytic cells, known as ***antigen-presenting cells*** (or ***APCs***), will "present" parts of the foreign antigen to ***inactive T lymphocytes.***

This presentation involves displaying of a ***major histocompatible protein (MHC)*** on the membrane of the phagocytic cell. Upon recognition of both the antigen and MHC, specific inactive T cells will become ***active T lymphocytes.*** The activated T cells will enlarge and secrete chemicals such as lymphokines. These chemicals will stimulate other immune cells, including ***helper T cells*** and ***cytotoxic*** (or ***killer***) ***T cells.*** The cytotoxic T cells are responsible for attacking and destroying any cell which displays the specific introduced antigen. The cytotoxic T cells combine with the foreign antigen and form the cell-antigen complex. The foreign antigen is then destroyed by cytolysis or by the action of macrophages. Helper T cells act as "directors," mediating the response of the cytotoxic T cells, suppressor T cells, and memory T cells. ***Memory T cells*** are created so that any future encounter with the specific antigen will ensure swift and immediate response by the T lymphocytes. ***Suppressor T cells*** act by slowing down and eventually ending the immune response, after the foreign antigen has been inactivated.

In addition, the helper T cells are required to ensure that inactive B lymphocytes become activated plasma cells. Therefore, if helper T cells are not produced, both cellular- and antibody-mediated immune responses are affected. The HIV virus, which causes AIDS, affects helper T cells and thus cripples the body's immune response.

Lab Activity

You will be watching 1–2 videos on the functioning of the immune system and answering questions concerning the videos (based on instructor preference).

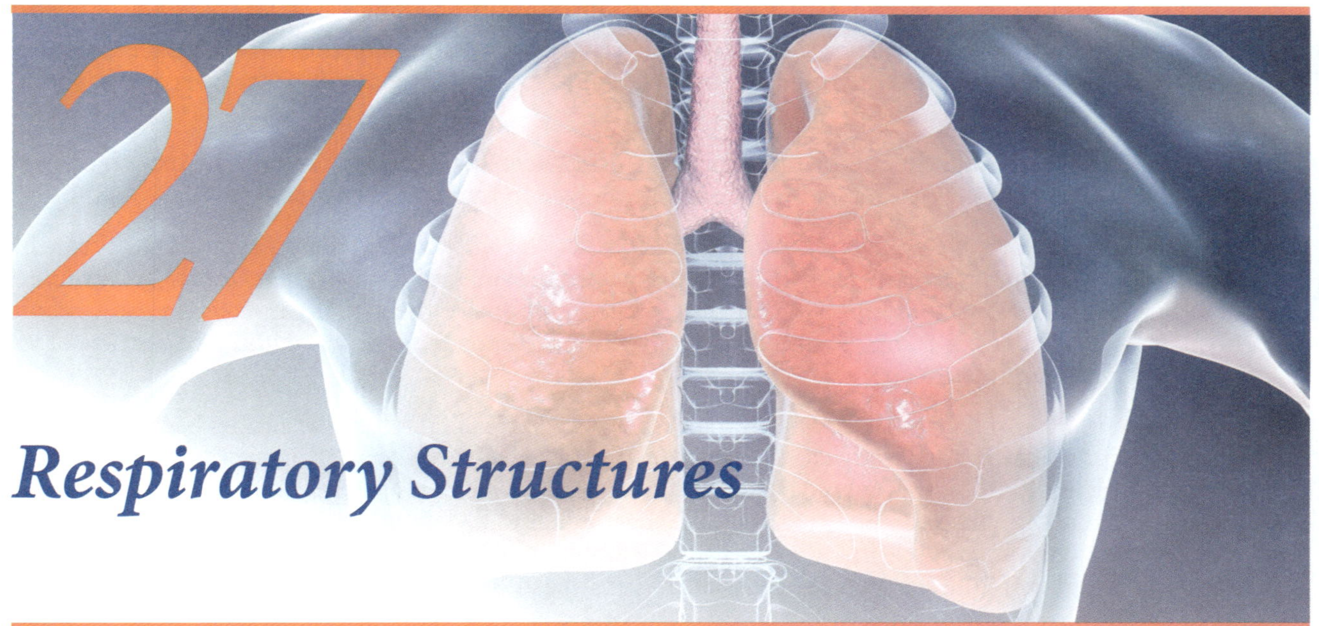

The respiratory passages are designed to deliver oxygen-rich air into and carbon dioxide out of the lungs. The respiratory system can be divided into two parts: the ***upper respiratory system*** which consists of the nose, pharynx and associated structures; and the ***lower respiratory system,*** containing the larynx, trachea, bronchi, and lungs.

The ***nose*** is divided into two chambers by the ***nasal septum.*** It is lined with mucus membranes and contains large numbers of blood vessels. Those structures allow the nose to warm, moisten, and filter the air.

At the back of the large nasal cavities, the air next passes through the ***internal nares*** (openings at the back of the nose) and then the ***pharynx*** (throat). The pharynx is divided into three parts: 1) The ***nasopharynx*** is the portion nearest the internal nares. It serves as a passage only for air. 2) The ***oropharynx*** is posterior to the oral cavity and extends inferiorly to the ***hyoid bone.*** This section is a passageway for both air and food/liquids from the mouth. 3) The ***laryngopharynx*** passes from the oropharynx to the ***larynx*** which is anterior and the ***esophagus*** posteriorly. This is also a passageway that serves both the respiratory and digestive systems.

The ***larynx*** is a highly cartilaginous structure, with nine pieces of cartilage composing its wall. Also within the larynx are the vocal cords, which is why the larynx is also known as the "voicebox." The anterior wall of the larynx is composed of a large, triangular-shaped cartilaginous structure known as the ***thyroid cartilage (Adam's apple).*** It tends to be larger in males due to the influence of testosterone.

The open space containing the vocal cords and other structures is known as the ***glottis.*** When swallowing, it is covered by a leaf-shaped piece of elastic cartilage called the ***epiglottis.*** This prevents food and liquids from entering the breathing passages ("going down the wrong tube").

At the base of the larynx is the ***cricoid cartilage,*** a ring of cartilage forming the inferior wall of the larynx where it joins the trachea. This piece of cartilage is used as a landmark for performing a tracheostomy to open up a blocked airway.

The ***trachea*** (windpipe) extends from the larynx, through the thoracic cavity towards the lungs. It is about 12" long and lies anterior to the esophagus. The tracheal walls contain "C" shaped rings of ***hyaline cartilage*** that are stacked on top of each other. The "open" portion of the "C" faces posteriorly so that the esophagus can expand slightly into the trachea when swallowing. Many organs of the lower respiratory tract also contain cartilaginous ring-like structures designed, in part, to ensure that the breathing passageways do not collapse and impede airflow.

The trachea divides inferiorly into the ***primary bronchi*** (right and left) which carry air towards the right and left lungs, respectively. At the point where the trachea divides into the bronchi, there is an internal ridge of highly sensitive tissue called the ***carina.*** If a foreign object touches the carina, it initiates a very strong cough reflex to expel the object.

The right bronchus is slightly more vertical, shorter, and wider than the left. Because of that structural difference, aspirated objects are more likely to become lodged in the right bronchus.

Upon entering the lungs, the primary bronchi divide into smaller and smaller tubes—from primary bronchi to ***secondary bronchi*** to ***tertiary bronchi*** and eventually to the smallest tubes—the ***terminal bronchioles.*** The bronchioles service the ***alveoli***—the air sacs in the lungs where the actual gas exchange will occur. This branching of tubes into smaller and smaller structures is known as the ***bronchial tree*** because of its resemblance to an upside-down tree.

Structurally, there are changes that occur in the tissues of the bronchial tree. As the tubes become smaller, the tissue gradually changes from ***ciliated pseudostratified columnar tissue*** containing many mucus-producing goblet cells to simple cuboidal cells in the terminal bronchioles. There is a shift in the amount of cartilage and smooth muscle in the walls, as well. There is a gradual loss of cartilage rings and an increase in the amount of smooth muscle in the walls. Without any supporting cartilage in the terminal bronchioles, a spasm of the smooth muscle, such as occurs during asthma attacks, can close off the airways.

The paired ***lungs*** are separated from each other by the heart, and other structures in the mediastinum. They are each surrounded by pleural membranes. The lungs extend from slightly above the clavicles to the diaphragm. On the medial aspect of the lungs is the ***hilus,*** a notched area where the pulmonary arteries, pulmonary veins, primary bronchi, nerves and lymphatic vessels enter and exit. Structurally, the left lung is longer and thinner than the right. The left lung is divided into 2 lobes, while the right lung has 3 lobes. ***Lobules*** are small segments in the lungs that each contain a lymphatic vessel, an arteriole, a venule, and a branch from a terminal bronchiole. At the terminal ends of these bronchioles are the ***alveoli,*** composed of simple squamous epithelial tissue for efficient and rapid gas exchange.

Required Respiratory Structures for Practical

After completion of the lab, you should be able to identify and correctly spell the following respiratory structures on the available models, pictures, and handouts.

Respiratory System Anatomy Model

1 | Nose

2 | Nasal cavity

3 | Nasopharynx

4 | Oropharynx

5 | Laryngopharynx

6 | Vocal cords

7 | Pulmonary arteries

8 | Pulmonary veins

9 | Right lung

10 | Left lung

Respiratory Tree Anatomy

1 | Hyoid bone

2 | Epiglottis

3 | Larynx

4 | Thyroid cartilage

5 | Cricoid cartilage

6 | Trachea

7 | Tracheal ring

8 | Carina

9 | Right primary bronchus

10 | Left primary bronchus

11 | Right secondary bronchus

12 | Left secondary bronchus

13 | Tertiary bronchus

Lobule Anatomy Model

1 | Pulmonary arteriole

2 | Pulmonary venule

3 | Terminal bronchiole

4 | Alveoli

5 | Alveolar sac

Histology

1 | Pseudostratified ciliated columnar epithelial tissue

2 | Hyaline cartilage (cartilaginous rings)

3 | Smooth muscle (trachealis muscle)

4 | Simple squamous epithelial tissue

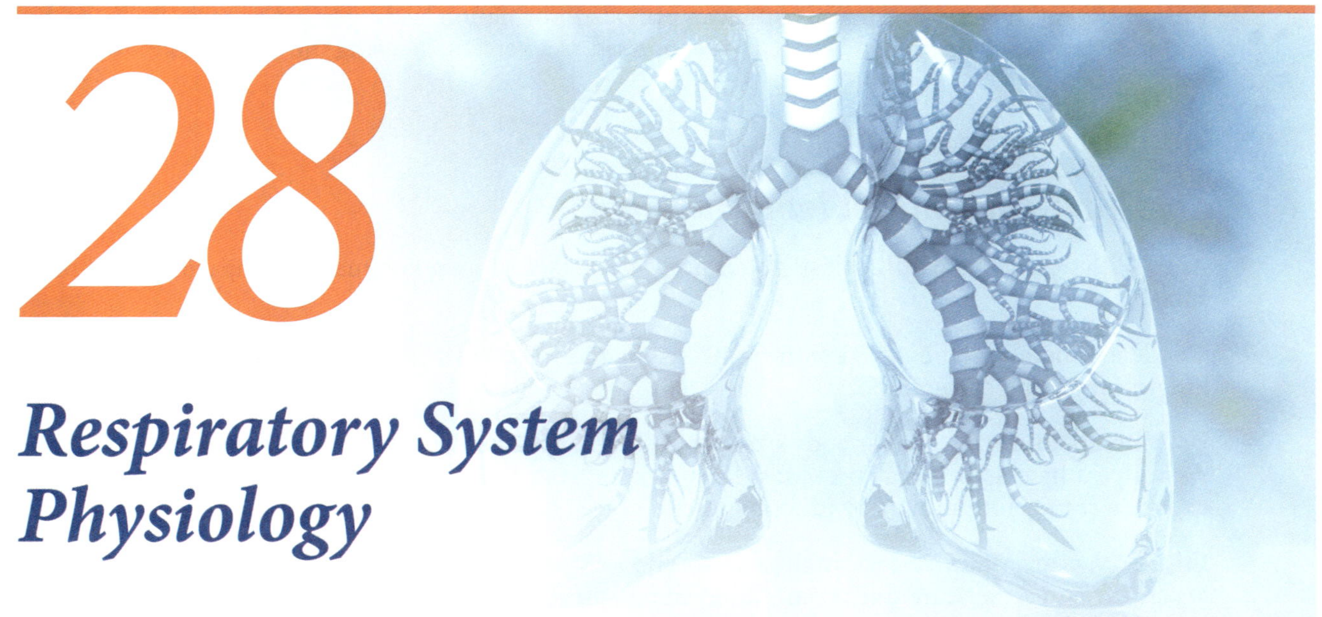

28 Respiratory System Physiology

Pulmonary Function Tests—Measuring Pulmonary Volumes and Capacities

Instruments known as ***spirometers*** are used to measure ***lung*** (or ***pulmonary***) ***volumes*** or ***capacities.*** Lung volumes are the amounts of air that can be moved into or out of the lungs. Pulmonary capacities can be calculated from the various pulmonary volumes. Knowing the various lung volumes and capacities are helpful in diagnosing lung disorders, in research, and in exercise training.

A normal respiratory cycle consists of ***inspiration (inhalation*** or ***breathing in)*** and ***expiration (exhalation*** or ***breathing out).*** This movement of air into and out of the lungs is known as ***pulmonary ventilation.*** The amount of air moved during normal pulmonary ventilation is known as the ***tidal volume.*** However, there is considerably more air that can be moved into and out of the lungs if needed. Below are some terms that you will need to know in order to complete this lab:

- ☐ **Tidal volume:** Amount of air inhaled or exhaled during a normal cycle of breathing.
- ☐ **Inspiratory reserve volume:** Amount of air which can be forcefully inhaled after a normal inhalation.
- ☐ **Expiratory reserve volume:** Amount of air that can be forcefully exhaled after a normal inhalation.
- ☐ **Vital capacity:** Amount of air that can be expired after a forceful inhalation. It can be calculated according to the formula:

  $$VC = TV + IRV + ERV$$

  You will be performing forced vital capacity (FVC) in lab and using that value in your calculations.
- ☐ **Forced expiratory volume in one second (FEV_1):** The volume of air that is forcibly exhaled from the lungs in the first one second of forced exhalation.
- ☐ **Residual volume:** The amount of air that remains in the respiratory tract after maximal exhalation. You can never completely empty the respiratory tract of air; there will always be a certain amount left in the respiratory tract organs.
- ☐ **Total lung capacity (TLC):** A value calculated according to the formula:

  $$TLC = VC + RV$$

- ☐ **FEV_1/FVC ratio:** Determining this ratio can provide information leading to a diagnosis of obstructive and restrictive lung disorders. This ratio is expressed as a percentage using the calculation:

  $(FEV_1 \div FVC) \times 100$

Factors That May Impact Lung Function

- **Age:** With aging, the natural elasticity of lungs decreases so lung volumes and capacities tend to get smaller.
- **Gender:** Generally, males who are the same height and weight of a comparably aged female have larger lung volumes and capacities.
- **Body height and size:** Smaller body sizes typically result in smaller values in pulmonary function tests. As body fat increases, it may prevent the diaphragm from fully functioning, resulting in smaller than expected results for pulmonary function tests.
- **Ethnicity:** Caucasians have the highest FVCs and FEV_1 values, with Mexican-American values slightly lower. African-Americans and Asians have values approximately 10–15% lower than Caucasians.

Uses for Pulmonary Functioning Testing

- Screening for the presence of certain obstructive and restrictive lung diseases or monitoring their progression.
- Evaluating a patient before surgery. This may be particularly important if the patient is over 60 years old, is obese, smokes, has asthma, will be under anesthesia for a long period, has a known pulmonary disorder, or is undergoing an abdominal or thoracic procedure.
- Determining how effective therapeutic intervention is.
- Monitoring a patient for ability to be weaned from a ventilator.

Pulmonary problems are often grouped into two categories—***obstructive*** and ***restrictive,*** based on two components that can be measured by a spirometer—airflow and volume of air.

If airflow is affected, the defect is usually ***obstructive.*** The openness of the air passages is estimated by measuring the flow of air as the patient exhales as quickly and hard as possible. With obstructive disorders, the person has difficulty breathing even while resting and finds expiration harder than inspiration.

If air volume is reduced, the problem is typically ***restrictive.*** The person with restrictive pulmonary disease will have a significantly reduced total lung capacity. Individuals with this problem usually have difficulty breathing when they exert themselves because their pulmonary ventilation (breathing) cannot keep up with increased demand.

Obstructed airflow can be due to the following:

- Asthma which causes smooth muscle of the bronchioles to contract, resulting in a narrowing of the airways.
- Bronchitis which causes narrowing of the airways from inflammation and swelling.
- Emphysema which destroys lung tissue and causes a loss of elasticity in lungs.
- Physical blockage of the passageways from excess mucus (such as occurs in cystic fibrosis), inhaling foreign objects, or a tumor.
- Airways being compressed by external trauma or tumors.

Restricted airflow is caused by lung disorders. It can be measured/determined by a loss of total lung capacity in patients. Causes of restricted airflow include the following:

- Disorders of the lungs such as tuberculosis or pneumonia.
- Factors outside of the lungs such as gross obesity, pregnancy, fluid in the pleural cavity, tumors, rib fractures, or pleurisy.
- Problems with the neuromuscular system, including paralysis of the diaphragm, muscular dystrophy, amyotrophic lateral sclerosis (Lou Gehrig's Disease), myasthenia gravis, polio, or generalized weakness.

Lab Activity

Using either wet or dry spirometers, you will measure various lung volumes and calculate lung capacities before and after exercise.

Materials Needed to Complete This Lab

- Wet or dry spirometer
- Disposable mouthpiece
- Noseclip, if available
- Calculator
- Stopwatch or clock with a second hand

A ***wet spirometer*** consists of a large plastic vessel inverted in a container of water. A breathing tube connects to the spirometer. While breathing through a mouthpiece and into the spirometer, the inverted container fills with air and will rise in the water. A numbered dial with a chain attached will provide a value for the amount of air displaced. This value will be recorded in the table at the end of the lab exercise.

A ***dry spirometer*** also measures the amount of air breathed into it. The dry spirometer is a hand-held device with a numbered dial on top, which provides the values for amount of exhaled air. When using the dry spirometer, it may be necessary to "zero" the scale at 1 liter in order for it to detect low volumes of air, such as the tidal volume. Be sure to take this into account when recording air values. As an example:

- Normal tidal volume is about 500 ml. Set the dial on 1 liter prior to breathing into the spirometer. After exhaling into the spirometer, the dial may read 1.5 liters. In order to figure the volume of air breathed into the machine, subtract 1 liter from 1.5 liters, which gives a value of .5 liters (or 500 ml)!

Prior to beginning the actual measurements, you may want to practice breathing in a normal way while pinching your nostrils or using nose clips, if available. Breathing through your nose while performing the measurements may give you inaccurate readings.

Lab Safety Note: The wet (and dry) spirometers used in the lab **do not** have one-way valves in them. **Do not inhale** through your mouthpieces or you may inhale potentially contaminated air or water.

Lab Safety Note: After your data have been collected, place your disposable mouthpiece in the waste container indicated by your instructor.

A | Measurement of Tidal Volume

1 | Zero the recording device. If using the wet spirometer, you must make sure that the rubber disc attached to the chain is set at "zero." If using the dry spirometer, note the zeroing technique indicated above.

2 | Place your disposable mouthpiece on the breathing tube of the wet (or dry) spirometer.

3 | With your nostrils pinched closed, ***inhale normally.***

4 | Place your mouth on the disposable mouthpiece and ***exhale normally*** into the tube.

5 | Read the result and record in Table 28.1. **Note:** Air volumes can be recorded as either liters or ml. The spirometers in the lab measure in liters, but some tables list values in ml. Please be consistent in how you record your values.

6 | Reset the dial to zero and repeat Steps 3–4 two additional times, recording the results in Table 28.1.

Lab Hint: There is actually very little air exhaled during a normal tidal volume (usually about 500 ml). A common mistake made with this measurement is breathing **too hard** into the machine. Try to breathe as normally as possible for accurate results.

B | Measurement of Expiratory Reserve Volume (ERV)

1 | Set the spirometer to zero.

2 | Before putting your mouth on the mouthpiece, ***inhale and exhale normally.***

3 | At the very end of normal exhalation, pinch your nostrils closed, place your mouth on the disposable mouthpiece, and ***forcefully exhale*** as much additional air as possible.

4 | Record your results in Table 28.1.

5 | Repeat Steps 1–3 two additional times and record your results in Table 28.1.

C | Measurement of Forced Vital Capacity (FVC)

1 | Set the recording device to zero.

2 | Before placing your mouth on the mouthpiece, ***inhale as forcefully*** as possible.

3 | With your nostrils pinched closed, immediately place your mouth on the disposable mouthpiece, and ***exhale forcefully*** as much additional air as possible.

4 | Record your result in Table 28.1.

5 | Reset the device to zero and repeat Steps 2–3 two additional times, recording your values in Table 28.1.

D | Measuring Forced Expiratory Volume (FEV_1)

Lab Note: For this portion of the lab, you will need a person who acts as a timer as well as a person who will record the results of the subject who is performing the test. The **timer** needs to be ready to call **"time"** at **exactly *one* second.** The recorder will take note of the volume indicated when the one-second time was called.

1 | Set the recording device to zero.

2 | Before placing your mouth on the mouthpiece, ***inhale as forcefully*** as possible.

3 | With your nostrils pinched closed, immediately place your mouth on the disposable mouthpiece and ***exhale forcefully*** as much additional air as possible until the timer calls "time" at 1 second.

4 | Record your result in Table 28.1.

5 | Reset the device to zero and repeat Steps 2–3 two additional times, recording your values in Table 28.1.

Lab Note: The spirometers we are using in lab will not provide the most accurate measurements for FEV_1, but should provide a reasonable approximation if you practice the technique a few times.

E | Calculating Expected Forced Expiratory Volume in One Second (FEV_1)

You will need to calculate your expected average FEV1 by using the formula below:

$$FEV_1 = FVC \times 0.8$$

F | Calculating Inspiratory Reserve Volume

We are not able to measure IRV using the spirometers. So this value will be calculated using the following formula:

$$IRV = VC - (RV + ERV)$$

Use your FVC value for the VC in the formula above.

Place your calculated IRV in Table 28.1.

G | Determining Residual Volume (RV)

Residual volume (RV) is the amount of air remaining in the respiratory tract after maximal exhalation. It cannot be measured. For calculations involving RV, use the following average values:

Average RV for adult male = 1.2 L

Average RV for adult female = 1.1 L

H | Calculating Total Lung Capacity (TLC)

To calculate Total Lung Capacity (TLC), use the numbers you obtained from your spirometer measurements and the following formula:

TLC = VC + RV

The value for RV should be 1.2 L for males and 1.1 L for females.

I | Calculating FEV_1/FVC Percentage

Calculating the FEV_1/FVC percentage value can provide valuable information in diagnosing obstructive and restrictive pulmonary disorders.

To calculate the percentage use the formula:

$(FEV_1 \div FVC) \times 100$

Results of Pulmonary Volume/Capacity Activities

In the table below, fill in *only* the white "squares." The blue shaded areas should not contain any numbers. Please Note: Use a pencil to record your results in Table 28.1.

Table 28.1 | Results of Pulmonary Volume/Capacity Activities

Volume or Capacity	Expected Value (included or obtained from charts)	Actual Value (obtained from lab activities or charts; each space represents one lab measurement.	Difference (between the expected value and your actual value)	Explanation What accounts for the difference? For each of the pulmonary tests you performed or for the calculations you made, indicate if your results were as you expected, better than expected, or worse than expected. If there were differences between the actual and expected values, please explain the reason(s) why.
Tidal Volume (TV)				
Average	*0.5 L (500 ml)*			
Expiratory Reserve Volume (ERV)				
Average	*1.1 L*			

Volume or Capacity	**Expected Value** (included or obtained from charts)	**Actual Value** (obtained from lab activities or charts; each space represents one lab measurement.	**Difference** (between the expected value and your actual value)	**Explanation** What accounts for the difference? For each of the pulmonary tests you performed or for the calculations you made, indicate if your results were as you expected, better than expected, or worse than expected. If there were differences between the actual and expected values, please explain the reason(s) why.
Vital Capacity (VC) (FVC)				
Average				
	Obtain the value for the square above from the tables included in lab specific for your age, sex, ethnicity, & height.			

Volume or Capacity	**Expected Value** (included or obtained from charts)	**Actual Value** (obtained from lab activities or charts; each space represents one lab measurement.	**Difference** (between the expected value and your actual value)	**Explanation** What accounts for the difference? For each of the pulmonary tests you performed or for the calculations you made, indicate if your results were as you expected, better than expected, or worse than expected. If there were differences between the actual and expected values, please explain the reason(s) why.
Forced Expiratory Volume in One Second (FEV_1)				
Average				
	For FEV_1 value above: FVC × 0.8			
Inspiratory Reserve Volume (IRV)				
Average				
	The value that goes in the square above neeeds to be calculated.	***The value that goes in the square above needs to be calculated.***		

Volume or Capacity	Expected Value (included or obtained from charts)	Actual Value (obtained from lab activities or charts; each space represents one lab measurement.	Difference (between the expected value and your actual value)	Explanation What accounts for the difference? For each of the pulmonary tests you performed or for the calculations you made, indicate if your results were as you expected, better than expected, or worse than expected. If there were differences between the actual and expected values, please explain the reason(s) why.
Total Lung Capacity (TLC)				
Average				
	Determine the expected value for the square above by using the nos. in this column and the equation: TLC = VC + RV.	*Determine the expected value for the square above by using the nos. in this column and the equation: TLC = VC + RV.*		
FEV_1/FVC Ratio *Calculate using the equation: (FEV_1 ÷ FVC × 100)*				

The values in the ***Expected Value*** column are those which would be expected from an average adult male or female. If those values are not included in the table, please consult the table included in this Lab Activity or in the lab.

The values for the ***Actual Value*** column are those that you measured with the spirometer or calculated using the formulas included in the lab. To calculate the average value, add the three ***Expected Value*** numbers together and then divide that number by three. For example, if your TV actual measurements are 0.8 L, 0.6 L, and 0.5 L, add those together for a sum of 1.8 L (0.7 + 0.5 + 0.6 = 1.8 L). Divide the 1.8 L by 3 to obtain your average TV: 1.8 L ÷ 3 = 0.6 L. So, your average TV is 0.6 L.

The ***Difference*** column should contain the difference between the Expected (Average) Value and your Actual Average Value.

The ***Explanation*** column should include a reasonable explanation as to why your Actual Value and Expected Values are different or the same.

Lab Note: Use the information below as well as the material contained in the introductory portions of this lab to help explain your results. (A website that is helpful for calculations: *cdc.gov/niosh/topics/spirometry/refcalculator.html* and *cdc.gov/niosh/topics/spirometry/nhanes.html.*)

Name: ____

Table 28.2 | Expected and Actual Values.

	What is "Normal"	Higher than Expected	Lower than Expected	Other
Forced Vital Capacity Usually this information by itself is inadequate to diagnose pulmonary function disorders.	Your actual value should be 80–85% of the expected value.	Individuals who were competitive athletes while their lungs were growing could be 120–130% above expected.	**Obstructive disease.** Usually need to monitor over a prolonged period of time. **Restrictive lung disease**. Because the amount of air that can be forcefully inhaled or exhaled is smaller due to the disease.	Using an inhaler to dilate the bronchial passages and then repeating the test can help distinguish if the issue is obstructive or restrictive. An improvement of 10–15% in FVC would suggest obstructive.
Forced Expiratory Volume in One Second	Healthy individuals should be able to expel 75–80% of their FVC in one second.		Highly indicative of obstructive disease.	
FEV_1/FVC Ratio	Should be 80–85% of expected value.		If this ratio is low in addition to a low FEV_1, it could indicate obstructive disease (emphysema, chronic asthma). Higher than expected FEV_1/FVC can be seen in fibrotic lung diseases (such as asbestosis).	

29

Respiratory System: Demonstration of CO_2 as a Waste Product

The process of exchanging gases (O_2 and CO_2) in the lungs is a passive process. However, if CO_2 is not eliminated and begins to accumulate, the body will utilize feedback mechanisms to try to rid itself of the extra CO_2.

The following chemical reactions occur in the body as CO_2 is produced. However, they can be replicated outside of the body as well.

$$\mathbf{CO_2 + H_2O \rightarrow H_2CO_3}$$

In the reaction above, adding CO_2 to H_2O produced an acidic compound carbonic acid (H_2CO_3).

The carbonic acid quickly dissociates to form a hydrogen and bicarbonate ion through the action of carbonic anhydrase:

$$\mathbf{H_2CO_3 \xrightarrow{\text{carbonic anhydrase}} H^+ + HCO_3^-}$$

This reaction results in an increased hydrogen ion concentration which will decrease the pH of the solution (make it more acidic). Recall that a pH below 7.0 is acidic, a pH of 7.0 is neutral, and a pH above 7.0 is alkaline or basic.

Adding sufficient hydrogen ions to a basic solution will change the pH from alkaline to acidic.

In this lab, we will demonstrate how your body's CO_2 production can change the pH of water.

Lab Activity

Materials Needed to Complete this Lab

- Beakers
- Distilled water
- Phenolphthalein
- Sodium hydroxide (NaOH)
- Clean drinking straw
- Parafilm
- Stopwatch

Procedure

1 | Fill a clean beaker about ½ full of distilled water.

2 | Add 2–3 drops of phenolphthalein. Phenolphthalein is used as a color indicator for a basic (alkaline) solution. It will turn an alkaline solution magenta in color (magenta is a bright reddish-purple color).

3 | Add a few drops of NaOH to make the solution basic. It should turn color with the addition of NaOH.

4 | Stretch a piece of Parafilm® across the top of the beaker and make sure it is securely attached to the sides of the beaker.

5 | Poke a hole through the Parafilm and insert your clean straw.

Lab Safety Note: Do not blow vigorously into the solution as it could cause the solution to splash onto your skin.

Lab Safety Note: After your data is collected, place your disposable straw in the waste container indicated by your instructor.

6 | While a lab partner keeps track of the time with a stopwatch, blow ***gently*** through the straw into the water solution until the color disappears. Record the amount of time it took for the color to disappear in Table 29.1. Keep your straw.

7 | Obtain a fresh beaker of distilled water and ***repeat Steps 2–5 above.*** Clear the stopwatch to zero. Do not blow into the water yet.

8 | Now, exercise vigorously until you are winded (2–3 minutes). Exercise could be running in place, doing jumping jacks, or running the stairs.

9 | As soon as you are winded, begin blowing through your straw into the solution. Make sure your lab partner is keeping track of the time with the stopwatch.

10 | Record your results in Table 29.1.

11 | Discard your used straw.

12 | Clean your beaker and place it on the drying rack above the lab sink.

Table 29.1 | Results

Result While at Rest	Result After Exercise
Time to turn the water from magenta to clear: ________________ seconds	Time to turn the water from magenta to clear: ________________ seconds

Answer the following questions:

1 | What chemical reaction caused the color to change from magenta to clear?

2 | After exercise, did the time to clear the water of color increase or decrease?

3 | How would you explain the difference that occurred after exercise?

If there is available time, you could repeat the tests comparing results from the following test subjects:

- male versus female
- smoker versus non-smoker
- athlete versus non-athlete
- asthmatic versus non-asthmatic

Record the results in Table 29.2:

Table 29.2 | Results

	Time to Turn the Water from Magenta to Clear (in seconds)
Male	
Female	
Smoker	
Non-smoker	
Athlete	
Non-athlete	
Asthmatic	
Non-asthmatic	

Answer the following questions:

1 | What difference(s) did you observe in each of the pairs in the table above (i.e., male vs. female; smoker vs. non-smoker, etc.)?

2 | Were any of the results surprising or unexpected?

3 | How could you explain these results?

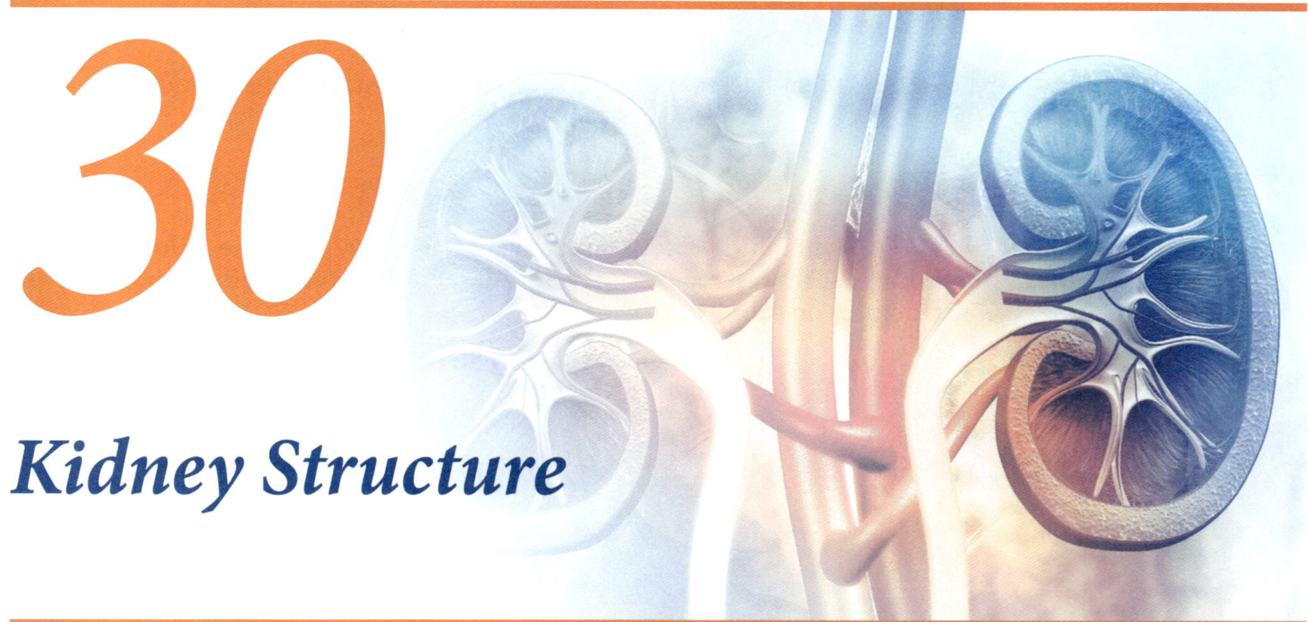

30 Kidney Structure

External Structure

The paired kidneys lie in the abdominal cavity, just above the waist area, and are held in place by a variety of structures. Structures that help to protect the kidneys and hold them in position include 1) the ***renal capsule*** which is a thin sheet of dense irregular connective tissue that covers the outer surface of the kidneys; 2) the ***adipose capsule*** which is fatty tissue that encases the renal capsule; and 3) the ***renal fascia*** composed of dense irregular connective tissue, which helps attach the kidneys to the abdominal wall.

At the medial, concave area of the kidney is the ***hilus,*** a notch through which the ***renal artery*** and ***renal vein*** enter and leave the kidney, and the ***ureter*** exits the kidney. In addition, nerves and lymphatic vessels associated with the kidney also can be found in this area.

Internal Structure

A frontal section of the kidney reveals the internal structures of the kidneys.

The outermost, reddish portion of the kidney is the ***renal cortex.*** The renal capsule surrounds the cortex. Internal to the cortex is a darker, reddish-brown area known as the ***renal medulla.*** In the cortex and medulla areas are millions of microscopic structures known as the ***nephrons,*** which are responsible for producing urine.

Included in the renal medulla are 8–10 triangular-shaped structures known as the ***renal pyramids.*** The broader ends (the ***bases***) of the pyramids face the cortex, while the narrower ends, the ***papillae,*** face the hilus. The renal pyramids are separated from each other by ***renal columns,*** which are extensions of the cortex material into the medulla area. The appearance of the columns is similar to that of the cortex.

As urine is formed in the nephrons, it drains into tubular ***papillary ducts*** that release the urine into cup-like structures, the ***calyces*** (***calyx*** = singular). The papillary ducts in each papilla drain the urine into ***minor calyces.*** The minor calyces merge into larger areas known as the ***major calyces.*** These, in turn, drain the urine into one large cavity known as the ***renal pelvis,*** which is near the hilus. From the renal pelvis, the urine exits the kidney through the ***ureter,*** a tubular structure that connects to the urinary bladder where urine is stored until it is eliminated from the body.

Lab Activity

Using your lab manual, textbook, and other references, locate the following structures on the kidney model.

Required Kidney Structures for Practical

After completion of the lab, you should be able to identify and correctly spell the following kidney structures on the available models. If preserved kidneys or cadaver kidneys are available, you may also be responsible for these structures on those specimens.

External Structures

- Hilus
- Renal artery
- Renal vein
- Ureter
- Renal capsule

Internal Structures

- Cortex
- Renal column
- Medulla
- Pyramids
- Base of pyramids
- Papilla of pyramids
- Minor calyces (calyx—singular)
- Major calyces (calyx—singular)
- Pelvis

31

Urinalysis Laboratory

The kidneys are responsible for a variety of physiological processes in the body. These include 1) regulating blood volume, blood pressure, and blood pH; 2) production of certain hormones; 3) excretion of waste products; and 4) regulating ion (electrolyte) levels in the blood. Urine production is the result of many of these processes. The urine is released from kidneys through the ureters, stored in the bladder, and eventually eliminated from the body through the urethra.

Normally the body releases approximately 1–2 liters of urine per day, although this volume can vary considerably based upon such factors as fluid intake, physical activity, and drugs like diuretics.

An individual's urine composition can also vary according to diet, physical condition, disease processes, and physical activity.

Urinalysis involves analyzing the volume, physical, chemical, and microscopic properties of urine. Such analyses can provide details about kidney function and help in diagnosing disease processes.

Lab Activity

Materials Needed to Complete This Lab

- Disposable gloves
- Test tubes and test tube holders
- Artificial urine specimens
- Urine dip sticks (such as Chemstrips®)
- Wax marking pencil
- Stopwatch or clock with a second hand

Working in groups of four, you will be analyzing some common physical and chemical components of artificial urine. Record your results in the table at the end of the exercise. To complete the table, your group will share its results with the other groups in the room. Each person is individually responsible for analyzing and diagnosing the results of each artificial urine sample.

A | Assessing the Physical Characteristics of Urine

The physical characteristics include visually assessing the color and transparency of urine.

The ***color*** of urine is normally a pale yellow to straw color. Color is primarily due to the presence of ***urochromes*** which are pigments produced during the breakdown of hemoglobin. Color can change based upon diet, drugs, or pathology.

Normal urine should be ***transparent*** (clear). Cloudiness can be a result of urine that has been standing too long or bacterial infections.

Odor can also be used to assess urine. Normally, urine odor should not be offensive. Ammonia-like smells can be a result of stale urine or from certain foods. Diseases can cause more offensive odors. An acetone-like odor can indicate diabetes. Odor can also vary based on diet and drugs.

Specific gravity is the measurement of density of a substance as compared to the density of pure (distilled) water. Pure water has a specific gravity of 1.000 (gm/ml). However, normal urine contains dissolved materials so its specific gravity will be greater than 1.000. Normal specific gravity for urine is 1.003–1.035 (gm/ml). This number can vary based upon the urine's fluid content and can also be indicative of certain diseases.

Lab Safety Note: Although we are utilizing artificial urine in the lab, assume that the material is a biohazard and wear disposable gloves. Place used gloves in the container indicated by your lab instructor.

1 | Obtain the artificial urine sample assigned to your group from the lab instructor.

2 | Using the wax pencil, label the outside of a test tube with the urine sample.

3 | Carefully fill the tube with about 25 ml of urine—this will fill the tube about half full—and place the tube in the tube holding rack.

4 | Examine the following physical characteristics of the tube:

- **Color:** In the table at the end of the exercise, describe the color of your urine sample. It is helpful to place a white piece of paper behind the sample to examine the color more accurately.
- **Transparency:** As you examine the color of the urine sample, also make note of the transparency (clear, slightly cloudy, cloudy, or very cloudy).

B | Assessing the Chemical and Other Constituents of Urine Using Reagent Test Strips

A variety of chemicals and other urine constituents, such as blood, can be assessed in urinalysis. The results can be important in determining many disease processes and kidney function.

Urine ***pH*** can range from 4.6–8.0. Most normal urine is slightly acidic, with a pH of about 5.0–6.0. pH measurements can vary based upon diet, becoming more alkaline with a vegetarian diet and some urinary tract infections, and more acidic with conditions such as starvation.

Under normal conditions, urine should not contain ***glucose.*** Trace amounts of glucose may appear after a meal that is high in carbohydrates. Higher amounts may indicate diabetes mellitus or pituitary disorders. The presence of glucose in urine is ***glucosuria.***

Protein, specifically albumin, should normally be present in trace amounts or not at all. Small quantities can occur after heavy exercise or due to certain diets. If the levels are abnormally high, it can be indicative of kidney infection, damage from high blood pressure, or damage to the kidneys from bacterial infections or heavy metals.

Ketones are chemicals produced during the metabolism of fats. They may appear in urine when an individual's diet is very low in carbohydrates, during fasting, or starvation. ***Ketonuria,*** the presence of ketones in urine, can also occur during uncontrolled diabetes mellitus.

Bilirubin, from the breakdown of hemoglobin, should not be present in normal urine or present in trace amounts. High levels, ***bilirubinuria,*** can occur with liver disease.

Blood is not normally found in urine. Blood in urine may appear with kidney damage, kidney stones, or urinary tract infections (UTIs). In addition, blood may occur in the urine sample of a menstruating female due to the proximity of the vaginal tract to the urethra, which may cause contamination of the urine with blood. Often urine is tested for the presence of ***hemoglobin*** rather than red blood cells as red blood cells are broken down in the production of urine.

Other constituents that might be tested in urinalysis include ***leukocytes*** (white blood cells). These may be present in high numbers with UTIs. ***Urobilinogen*** is a byproduct of hemoglobin destruction. More than trace amounts of urobilinogen can occur with liver disease or excess red blood cell destruction.

To test for these and other components in urine, reagent test strips are used. The test strips have small squares of specially prepared reagent pads that will change color when exposed to the various constituents in urine. These are compared to color charts on the outside of the container to obtain the lab values for each constituent.

Lab Note: Dependent upon the type of strips available in the lab, you may not be testing for all of the components discussed above. Examine the package to determine which components your specific reagent strips measure.

1 | Obtain a package of lab reagent strips.

2 | You will use the same urine sample contained in the tube you used for physical analysis.

3 | Prior to beginning the analysis, notice the information on the outside of the reagent container specifying the time intervals for reading each urine constituent (e.g., some values are "read" or compared to the color chart after 30 seconds and some after 60 seconds).

Lab Note: Timing of the readings is critical.

Assign one individual in the group as the "timer" who will indicate when the critical time intervals occur.

4 | Quickly dip **one** reagent test strip briefly in the urine sample, ensuring that all of the pads are wet. You should not hold the test strip in the urine for more than about one second.

5 | After removing the strip from the sample, tap excess urine on the edge of the tube. Be careful not to allow the reagents to run together. ***The timer should begin timing at this point.***

Lab Note: Do not dip the strip more than one time and use only **one** strip for all of the analyses of your urine sample.

6 | Hold the test strip vertically and then, at the indicated time intervals, compare each pad to the specific color charts on the outside of the container. In comparing colors, hold the reagent strip closely to the color chart.

7 | Record your results in Table 31.2.

8 | Dispose of the artificial urine and test strips according to directions given by your lab instructor. You will also need to thoroughly wash and rinse your test tubes and place them upside-down to dry.

9 | Share the results from your lab sample with the other lab groups.

C | Microscopic Examination of Urine

Urine contains a variety of microscopic materials that are also important in diagnosis. Some of these materials include the following:

- ***Epithelial cells*** that are sloughed from the lining of the walls of the urinary tract; the presence of these is normal.
- A variety of ***blood cells,*** including white blood cells which may indicate infection.
- ***Crystals*** of various chemicals that are normal components of urine or crystals formed from drugs. If these crystals form large masses, they are known as ***calculi*** or ***stones.***
- ***Casts,*** which are hardened masses of materials that have sloughed off into the urine. They may include casts of epithelial cells, blood cells, or bacteria. The casts resemble the materials from which they originated. Some casts are normal but others are not.
- ***Artifacts,*** which are materials that have inadvertently gotten into the urine sample. Artifacts can include clothing fabric, sperm, or even oil from the skin.

1 | Because the lab utilizes artificial urine, none of the materials discussed above will be present in the lab specimens.

2 | However, there are lab manuals available that illustrate some of the materials mentioned above. Please examine the manuals prior to finishing the urinalysis lab.

Use the information in the table below to help you analyze your lab results.

Table 31.1 | Urinalysis Lab Information

Laboratory Test	Normal Value	Abnormal Values and Significance
Color	Light yellow to amber	**Pale yellow:** Dilute urine from ingestion of large amount of water or diabetes insipidus (ADH disorder). **Dark yellow-dark amber:** Concentrated urine from dehydration or excess ADH, heart failure. **Reddish:** Consumption of beets or pigments from other foods, some medications, or blood. **Greenish:** Bile pigments from jaundice, some bacterial infections. **Brown to brownish-black:** Poisoning from heavy metals, hemorrhage from physical damage to kidneys; some medications (e.g., levodopa, flagyl). **Milky white:** Presence of fat or pus (white blood cells) which would indicate urinary tract infection.
Transparency	Clear	Cloudiness can occur when urine is "stale" (has been standing too long). Excess cloudiness can be pus in urine from infection.
Leukocytes	Zero	Trace or above may indicate a urinary tract infection (UTI).
Nitrites	Negative	Presence of nitrites may indicate certain bacteria and a possible urinary tract infection.
Urobilinogen	None or trace amount	High levels may occur with liver disease (infectious hepatitis, cirrhosis) or excessive destruction of red blood cells, jaundice, certain anemias (pernicious, hemolytic), and congestive heart failure.
Protein (albumin)	None or trace (no more than 10 mg/100 ml in a single sample or <100 mg/24 hr of urine collection); +1 = positive	Trace amounts with certain diets (high protein), pregnancy, or strenuous exercise regimens. Higher levels (>250 mg/day) may be due to kidney infection or inflammation, hypertension, damage from NSAIDS, heavy metal damage, or heart failure.
pH	Average of 5.5–6.5 (range 4.6–8.0)	More alkaline (higher pH) with vegetarian diet and some urinary tract infections. More acidic in starvation, fasting, athletes, and with acidic diet.
Blood (hemoglobin)	None	If present, may indicate kidney trauma, disease, or tumor. Inflammation of other urinary tract organs from kidney stones. Can appear in urinary tract infections. Can be a result of severe burns. May be contamination from female menstrual blood.
Specific Gravity	1.003–1.035 (gm/ml)	Higher when urine is concentrated; lower when urine is dilute. High values can be due to diabetes mellitus, dehydration, increases in ADH production. Low values occur in diabetes insipidus or renal failure.
Ketones	None	May appear in urine due to fasting, starvation, diet low in carbohydrates, incomplete fat metabolism, uncontrolled or poorly controlled diabetes mellitus.
Bilirubin	None or trace amount	High levels can occur with liver disease such as hepatitis or cirrhosis.
Glucose	None	Trace may appear after a high carbohydrate meal. Higher levels can result from diabetes mellitus, pituitary problems, liver disease, or stress.

Results of Urinalysis Testing

Note: Dependent upon the type of reagent dipstick available in lab, you may not have results for some of the chemicals indicated below. Please refer to your particular package or instructor. Place the results from your specific urine sample in the correct column, and then obtain results for the other urine samples from the other lab groups.

Name: ______________________________

Table 31.2 | Results of Urinalysis Testing

Lab Test	Sample A	Sample B	Sample C	Sample D	Sample E	Sample F
Color						
Transparency						
Leukocytes						
Nitrites						
Urobilinogen						
Protein						
pH						
Blood						
Specific Gravity						
Ketones						
Bilirubin						
Glucose						

After completing Table 31.2, utilize the information provided in Table 31.1 to evaluate each of the urine samples. For each test, you will need to determine if the lab value is normal or abnormal and complete Table 31.3. In the Discussion/Diagnoses column of Table 31.3, discuss what conditions may have produced any abnormal value. If all lab values were normal, you would place "normal" in that column.

Name: ______________________________

Table 31.3 | Results of Urinalysis Testing: Normal or Abnormal Values and Diagnoses

	Lab Tests Which Show Normal Values	Lab Tests Which Show Abnormal Values	Discussion/Diagnoses
Sample A			
Sample B			
Sample C			
Sample D			
Sample E			
Sample F			

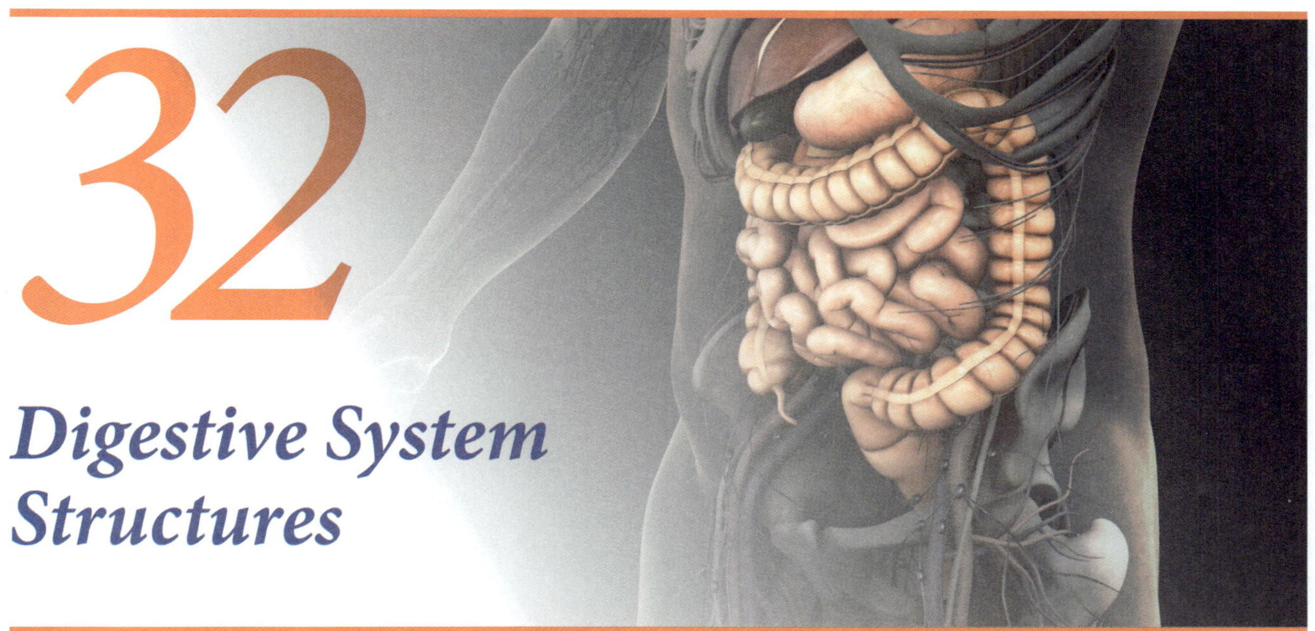

32 Digestive System Structures

After completion of the lab, you should be able to identify and correctly spell the following digestive system structures on the available models.

Required Digestive System Structures for Practical

Head Portion of Digestive Model

- Mouth (oral cavity)
- Teeth
- Tongue
- Parotid salivary gland
- Pharynx (oropharynx)
- Esophagus

Stomach and Gallbladder Model

- Esophagus
- Gallbladder
- Fundus of stomach
- Longitudinal muscle layer
- Circular muscle layer
- Oblique muscle layer
- Gastric rugae

- Pylorus (pyloric sphincter)
- Duodenum

Abdominopelvic Organs Model

- Stomach
- Liver
- Pancreas
- Jejunum
- Ileum
- Ileocecal valve
- Cecum
- Ascending colon
- Transverse colon
- Descending colon
- Sigmoid colon
- Rectum

Intestinal Villi Model

- Villi
- Simple columnar epithelial tissue
- Central lacteal
- Arteriole
- Venule
- Capillary network

33

Chemical Digestion

The purpose of the digestive system is to break down foods into components that can be absorbed. There are several parts to the digestive process:

1 | **Ingestion:** Taking the food materials into the body.

2 | **Peristalsis:** Movement of the materials through the organs; this involves muscle contraction within the digestive organs.

3 | **Mechanical digestion:** The physical breakdown of larger food materials into smaller components.

4 | **Chemical digestion:** The chemical breakdown of food materials into components that can be absorbed. These processes involve the use of enzymes.

5 | **Absorption:** Movement of the end products of chemical digestion into the blood for transport throughout the body. Both the cardiovascular and lymphatic systems are involved in absorption.

6 | **Defecation:** Elimination of the non-digestible components.

Mechanical digestion involves physical processes (e.g., chewing, churning of food in the stomach, mixing with bile salts) that increase the surface area of ingested materials by changing their physical form. This increase in surface area makes chemical digestion more efficient. The use of bile salts in this lab will illustrate how physical alteration in lipids influences lipid digestion.

Chemical digestion involves the use of specific enzymes which drive chemical hydrolysis and dehydration synthesis reactions. These reactions break down organic molecules (carbohydrates, lipids, proteins, nucleic acids) into their simplest forms which can be absorbed in the small intestine. Our body's enzymes have a range of conditions where their performance is optimal. Temperature and pH are two factors that impact enzyme activity. In this lab, we will observe how alterations in temperature affects enzyme activity.

Materials Needed to Complete this Lab

Chemicals/Reagents:

- Distilled or deionized water
- Pancreatin powder (and spatula)
- Bile salt powder (and spatula)
- Benedict's reagent
- Iodine reagent (such as Lugol's)
- Litmus cream
- 1% starch solution
- 1% glucose solution
- Vegetable oil (dyed with Sudan 4)

Lab Materials:

- 16 test tubes (2 test tubes with caps)
- Test tube rack and tongs
- Wax pencil
- Disposable 1-ml measuring droppers
- Ice water bath (with test tube rack)
- Warm water bath (37°C with tube rack)
- Hot plate
- Beakers (500-ml)

Lab Note: Read through the entire lab prior to starting the procedures as there are different time spans to complete the experiments. Some experiments take much longer than others and should be started first! Assign a lab partner to keep track of all times!

First things first! Using the wax pencil, label your test tubes prior to beginning the lab. Label the two test tubes with caps as 2a and 2b. Label the remaining test tubes as 1a, 1b, 1c, 3a, 3b, 3c, 4a, 4b, 5a, 5b, 6a, 6b, 7a, and 7b.

Lab Activity 1 | Lipid Digestion

Lipids are polymers of glycerol and fatty acids. To breakdown lipids into the monomers, both mechanical and chemical digestion is involved. Bile salts (secreted by the liver) are used in mechanical digestion of lipids through **emulsification.** Emulsification involves breaking up large lipid droplets into smaller ones; this produces more surface area on which the enzyme **lipase** can react to produce the monomers through chemical digestion. **Pancreatin** (secreted by the pancreas) contains many digestive enzymes, including pancreatic lipase.

In this portion of the lab, you will be using litmus cream (cream with litmus added) and observing the effects of bile and pancreatic lipase on lipid digestion. Litmus is a color indicator that turns from blue to red (or pink) in the presence of acids (such as fatty acids).

Materials Needed For Activity 1

- Test tubes labelled 1a, 1b, 1c, 2a, 2b
- Disposable droppers
- Distilled water
- Vegetable oil
- Bile salts and spatula
- Pancreatin and spatula
- Litmus cream
- Warm water bath

A | Procedure for Observing Lipid Digestion by Pancreatic Lipase

1 | Using disposable droppers, place the following materials in test tube **1a.**

- a | 2 ml of litmus cream
- b | 2 ml of distilled water
- c | This tube will be the control tube.

2 | Using disposable droppers or spatula, place the following materials in test tube **1b.**

- a | 2 ml of litmus cream (with dropper)
- b | 2 ml of distilled water (with dropper)
- c | Small amount of pancreatin powder (with spatula); you will need just enough powder to cover the end of the spatula.

3 | Using disposable droppers or spatulas, place the following materials in test tube **1c.**

- a | 2 ml of litmus cream (with dropper)
- b | 2 ml of distilled water (with dropper)
- c | Small amount of pancreatin powder (with spatula); you will need just enough powder to cover the end of the spatula.
- d | Small amount of bile salts (with new spatula); you will need just enough powder to cover the end of the spatula.

Please Note: Utilize the hood or wear a mask when adding bile salts to the tube.

4 | Place all of three tubes into the warm water bath (37°C) and observe your tubes periodically for about **an hour (check your tubes every 15 minutes).** Put your results in **Table 33.1.**

5 | Note the following:

- a | Time placed in the water bath: ______________________________
- b | Time removed from the water bath: ______________________________
- c | If fatty acids form, the litmus cream will turn from its original blue tint to lavender-pink, to pink over the hour time frame.

6 | Predict Questions:

- a | What do you predict will occur with test tube 1a? Will you see any color changes? Will there be a difference over time?

b | What do you predict will occur with test tube 1b? Will you see any color changes? Will there be a difference over time?

c | What do you predict will occur with test tube 1c? Will you see any color changes? Will there be a difference over time?

Table 33.1 | Lipid Digestion

Tube	Initial Observation	Observation after 15 Minutes	Observation after 30 Minutes	Observation after 45 Minutes	Observation after 60 Minutes
1a: Litmus cream and distilled water					
1b: Litmus cream, distilled water, and pancreatin					
1c: Litmus cream, distilled water, pancreatin, and bile salts					

B | Procedure for Observing the Emulsification Effect of Bile

1 | Using disposable droppers, place the following materials in test tube **2a (with the cap).**

a | 2 ml of distilled water

b | 2 ml of vegetable oil

c | This tube will be the control tube.

2 | Using disposable droppers, place the following materials in test tube **2b (with the cap).**

a | 2 ml of distilled water

b | 2 ml of vegetable oil

c | Small amount of bile salts; you will need just enough powder to cover the end of the spatula.

Please Note: Utilize the hood or wear a mask when adding bile salts to the tube.

3 | Cap both tubes and shake them **simultaneously** for about 30 seconds.

4 | Compare the distribution of oil in both tubes.

5 | Continue watching them for about **10 minutes.**

6 | **Predict Questions:**

a | What do you predict will occur with test tube 2a? Will there be a difference over time? Explain.

b | What do you predict will occur with test tube 2b? Will there be a difference over time? Explain.

7 | Record your results in **Table 33.2.**

Table 33.2 | Lipid Digestion

Tube	Initial Observation	Observation after 5 Minutes	Observation after 10 Minutes
2a: Veg. oil and distilled water			
2b: Veg. oil, distilled water, and bile salts			

Lab Activity 2 | Carbohydrate Digestion

Carbohydrates are polymers of saccharides. There are a variety of enzymes that will break down carbohydrates into their smallest components (monosaccharides). These enzymes include **amylase** which is produced in both the salivary glands (salivary amylase) and pancreas (pancreatic amylase.)

In this portion of the lab, you will be using the large polysaccharide starch and examine the role of pancreatic amylase (contained in pancreatin) and temperature on the digestion of starch. To test for the presence of starch and its sugar products, you will be using an iodine solution (to test for the presence of starch) and Benedict's reagent (to test for the presence of simple sugars, such as glucose).

If starch is present, the iodine solution will turn from its normal dark yellow color to blue-black. If there is no color change (if it remains dark yellow), then that indicates a negative result (meaning that starch is **not** present).

Materials Needed For Activity 2

- Test tubes labelled 3a, 3b, 3c, 4a, 4b, 5a, 5b, 6a, 6b, 7a, 7b
- Disposable droppers
- Distilled water
- 1% glucose solution
- Starch solution
- Iodine indicator
- Benedict's reagent
- Pancreatin and spatula
- Hot plate with beaker of boiling water
- Tongs
- Warm water bath
- Ice water bath

A | Procedure for Analyzing Whether a Solution Contains Starch and Its Sugar Products

1 | Using disposable droppers, place the following materials in test tube **3a.**

- a | 2 ml of starch solution
- b | A few drops of the iodine indicator
- c | Note the color in **Table 33.3.**
- d | This tube will be the control tube.

2 | Using disposable droppers, place the following materials in test tube **3b.**

- a | 2 ml of distilled water
- b | 2 ml of Benedict's reagent
- c | This tube will be the control tube to test for the presence of sugars.

3 | Using disposable droppers, place the following materials in test tube **3c.**

- a | 2 ml of distilled water
- b | 2 ml of 1% glucose solution
- c | 2 ml of Benedict's reagent

4 | Place tubes **3b** and **3c** in a beaker of boiling water (on a hot plate) for **3 minutes.**

> **Please Note: Use the tongs to carefully remove the test tubes from the boiling water at the end of the 3 minutes.**

- a | Note the following:
  - i | Time placed in the boiling water: ______________________________
  - ii | Time removed from the boiling water: ______________________________

5 | Note any color changes. Tube 3b should show a negative result—remain blue; tube 3c should show a positive result—a change from blue to another color (such as yellow, orange, or red). Put your results in **Table 33.3.** Keep these tubes for later reference!

Table 33.3 | Analyzing for Starch and Sugar

Tube	Initial Observation	Observation after 3 Minutes
3a: Starch solution and iodine indicator		
3b: Distilled water and Benedict's reagent		
3c: Distilled water, glucose solution, and Benedict's reagent		

B | Procedure for Examining the Effect of Temperature on the Digestion of Carbohydrates

1 | Using disposable droppers, place 2 ml of starch solution in tubes **4a, 4b, 5a, 5b, 6a, 6b, 7a,** and **7b.**

2 | In those same test tubes, simultaneously add the following using the spatula:

a | Small amount of pancreatin; you will need just enough powder to cover the end of the spatula.

3 | **Immediately:**

a | Place **4a** and **4b** in an ice water bath (0–1°C).

i | Time placed in the ice water bath: ______________________

ii | Time removed from the ice water bath: ______________________

b | Place **5a** and **5b** in a rack at room temperature (20–25°C).

c | Place **6a** and **6b** in a warm water bath (37°C).

i | Time placed in the warm water bath: ______________________

ii | Time removed from the warm water bath: ______________________

d | Place **7a** and **7b** in a boiling water bath (100°C).

i | Time placed in the boiling water: ______________________

ii | Time removed from the boiling water: ______________________

4 | Remove the tubes from the respective water baths **after 30 minutes.** (If the tubes in the boiling water bath decrease to 0.5 ml or less, remove them earlier.)

Please Note: Use the tongs to carefully remove the test tubes from the boiling water at the end of the 30 minutes.

5 | While the tubes are processing, answer the following predict questions:

a | Predict what role temperature may play in carbohydrate (starch) digestion.

b | Predict which test tubes will show the most enzymatic digestion from starch to sugars. Explain your prediction.

c | Predict which test tubes will show the least enzymatic digestion from starch to sugars. Explain your prediction.

d | What do you predict will happen over time?

6 | At the end of 30 minutes, put all of the test tubes in a test tube rack.

7 | For test tubes **4a, 5a, 6a,** and **7a:**

a | Using a disposable dropper, place a few drops of the iodine indicator into each of these test tubes.

b | This will test for the presence of starch. Refer back to Table 33.3, tube 3a, to determine if starch is present in these tubes.

c | Put your results in Table 33.4.

8 | For test tubes **4b, 5b, 6b,** and **7b:**

a | Using a disposable dropper, place 2 ml of Benedict's reagent in each tube.

b | Then **boil these tubes for 3 minutes.**

c | This will test for the presence of sugar.

d | Note the following:

i | Time placed in the boiling water: ______________________________

ii | Time removed from the boiling water: ______________________________

Please Note: Use the tongs to carefully remove the test tubes from the boiling water at the end of the 3 minutes.

e | Refer back to Table 33.3, to determine if sugar is present in these tubes.

f | Put your results in Table 33.4.

Table 33.4 | Carbohydrate Digestion

Tube	Initial Observation	Observation After 30 Minutes	Result of Starch Test (positive or negative)	Result of Sugar Test (positive or negative)
4a (ice water bath): Starch solution, pancreatin, and iodine indicator				
5a (room temperature): Starch solution, pancreatin, and iodine indicator				
6a (warm water bath): Starch solution, pancreatin, and iodine indicator				
7a (boiling water bath): Starch solution, pancreatin, and iodine indicator				
4b (ice water bath): Starch solution, pancreatin, Benedict's reagent				
5b (room temperature): Starch solution, pancreatin, and Benedict's reagent				
6b (warm water bath): Starch solution, pancreatin, and Benedict's reagent				
7b (boiling water bath): Starch solution, pancreatin, and Benedict's reagent				

Laboratory 33 Review Questions

Chemical Digestion

Name: ____

Analysis of Your Results

Lipid Digestion

1 | Did your results match your predictions? Explain.

2 | Did bile salts influence the digestion of lipids? Explain.

3 | Did any of the tubes turn pink? How long did it take? Explain the relationship.

Carbohydrate Digestion

1 | Did starch digestion occur in any of the test tubes? Which tubes? How did you determine if starch digestion occurred?

2 | Did temperature affect carbohydrate digestion? If so, how?

3 | Did your predictions on temperature effects match your results? Explain.

4 | Which temperature showed the most digestion? The least? Relate this information to your own physiology.

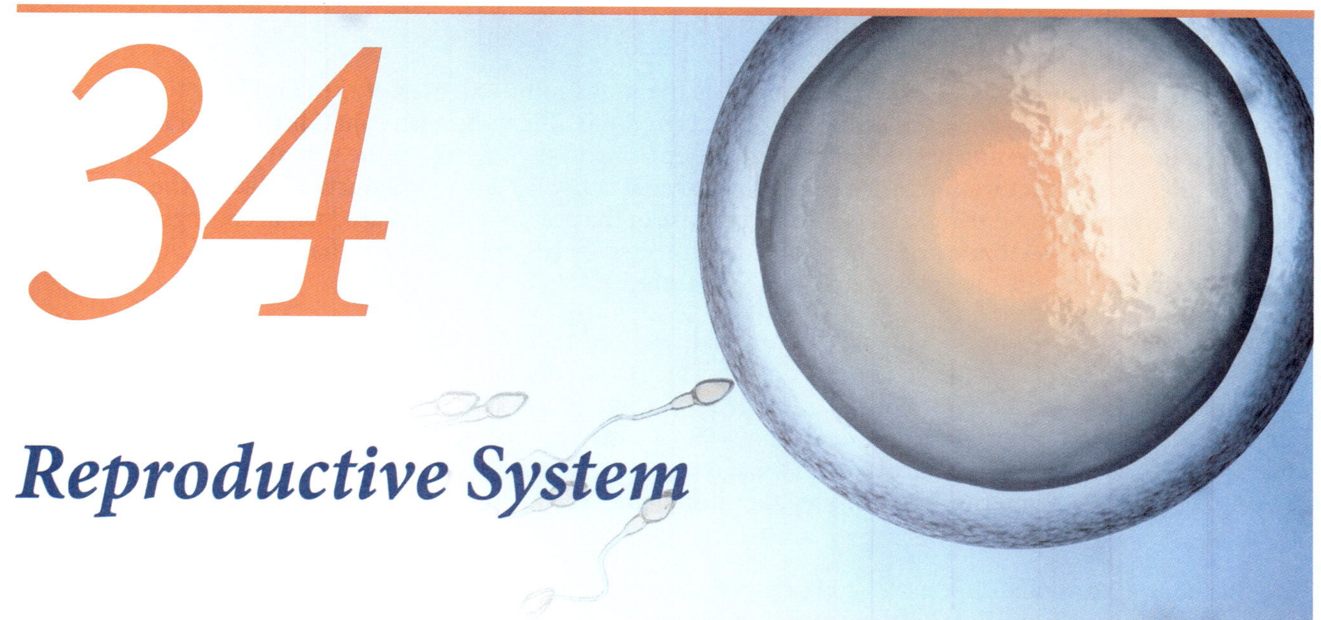

34 Reproductive System

Reproductive System Overview

The reproductive system is different from the other organ systems in the body. Whereas the other body systems are designed to maintain homeostasis, the reproductive system's primary function is to ensure survival of the species. It also promotes genetic diversity and allows a species to adapt to and survive in its environment.

To accomplish the primary function, the assigned male and female reproductive systems produce cells called **gametes.** Gametes are unique from other body cells by the manner in which they are produced and by the total number of chromosomes they contain. The cell division process to produce the gametes is called **meiosis.**

Gamete Production

You may recall that mitosis produces two new cells each containing a total of 46 chromosomes (23 pairs). The term **diploid (2*n*)** is used to describe the cells created by mitosis. In contrast, the process of meiosis results in four new cells each of which contains only 23 chromosomes; these cells are referred to as **haploid (*n*).**

The locations of gamete production are the **gonads.** In the male reproductive system, the gonad is called the **testis** (singular) **(testes** = plural). The process of producing sperm is known as **spermatogenesis.** In people assigned male at birth and having functional testes, spermatogenesis begins at puberty and continues for the rest of their lives.

In the female reproductive system, the gonad is the **ovary.** The gamete produced in the ovary is called an **ovum** (singular). The process of producing **ova** (plural) is known as **oogenesis** which begins during fetal development. There are two structures produced during oogenesis. One structure is the **oocyte.** The other structure is the follicle, which is a group of cells surrounding the developing oocyte. The process of oogenesis is halted before birth and then resumed with the onset of puberty, in individuals assigned female at birth with functional ovaries. The process of producing an **oocyte** continues each month until the individual reaches menopause, when the process is terminated. Only if an oocyte is fertilized will it become the fully functioning **ovum.**

Female Reproductive Anatomy Overview

Externally, the female's genitalia (or **vulva**) includes the **mons pubis,** an area of skin and fatty tissue, covered with pubic hair, overlying and cushioning the pubic symphysis. The **vulva** also contains the **labia majora, labia minora, clitoris,** the opening to the vagina, and the opening to the urethra of the urinary system. The **clitoris** is an erectile tissue structure contained inside the vulva; it is not used for reproduction. Rather the function of the clitoris is for pleasure during sexual activity because of the presence of increased nerve endings. The vaginal opening is called the **vestibule** and is covered and protected by two folds of skin. The external folds of skin are the **labia majora,** and the inner folds are the **labia minora.**

The **vagina** is the elastic, muscular part of the female genital tract. It extends from the vestibule to the cervix. Functions of the vagina include acting as an exit for menstrual discharge, serving as the lower end of the birth canal for childbirth, and accommodating the penis during sexual intercourse. The **cervix** is the lowermost, narrow end of the uterus that forms a canal between the uterus and vagina. The cervix acts as a doorway for sperm to pass through on the journey to fertilize the oocyte. During pregnancy the cervix's role is to keep the developing fetus in the uterus. In a nonpregnant individual, the cervix prevents pathogens from invading the uterus.

The **uterus** is a hormone responsive, cone-shaped, hollow muscular organ located in the pelvis between the bladder and the rectum. The inferior tapered portion of the uterus is the cervix, and the superior broader portion is the **fundus.** Extending from the lateral portions of the uterus is the **broad ligament** that helps hold the uterus in place by anchoring it to the lateral walls of the pelvic cavity. Additionally, the uterus is anchored by the round ligament, which is 10–12 inches in length extending from the superior lateral wall of the uterus towards the inguinal region.

Extending from the fundus of the uterus to the ovaries are the four-inch long right and left **fallopian tubes (uterine tubes; oviducts).** Their function is to transport the oocyte from the ovaries to the uterus. At the superior end of the fallopian tubes, the structure changes into finger-like projections called **fimbriae.** During ovulation the fimbriae become engorged with blood and gently brush across the surface of the ovary to pick up the ovulated oocyte.

The paired **ovaries** are the gonads of the female reproductive system. Each ovary is held in position by the **ovarian ligament** which extends from the superior lateral surface of the uterus (right and left) to the ovary. Ovaries are responsive to **follicle stimulating hormone** and **luteinizing hormone** produced by the pituitary gland. These two hormones stimulate the final part of oogenesis and induce ovulation of the mature oocyte. Additionally, the ovary produces two hormones—**estrogen** and **progesterone.** Both of these hormones function to prepare the uterus for implantation of a fertilized egg.

Male Reproductive Anatomy Overview

The external male reproductive anatomy also includes the mons pubis. However, the mons pubis in the male is less developed than a female's due to the lack of the hormone estrogen. At the inferior aspect of the mons pubis is the **penis** which is the sexual organ of the male reproductive anatomy. It is also an organ of the urinary system as urine exits the body through the urethra within the penis. The penis consists of several parts: the **root,** the **body (shaft),** and the **glans penis.** The **root** attaches the penis to the body wall and extends into the pelvic cavity. The elongated **body (shaft)** contains two types of internal erectile tissue, the **corpora cavernosa** and **corpora spongiosum.** During sexual arousal these two tissues become engorged with blood causing the penis to become erect. At the distal end of the body is an enlarged tip—the **glans penis (glans).** The glans surrounds the **urethral orifice (meatus)** which is the external opening of the urethra used for urination and ejaculation. Some males (non-circumcised) may have a **prepuce** (foreskin) covering the glans penis. The prepuce protects the glans from injury and secretes lubricants to protect the glans from infection.

The **scrotum (scrotal sac)** is also part of the male external genitalia; it is located inferiorly to the penis. Internally, the scrotum is divided into two compartments, each of which contains a **testis.** Besides being the site of spermatogenesis, the testes also produce most of the hormone testosterone after the individual reaches puberty. There are two muscles associated with the scrotum: the **cremaster muscle** composed of skeletal muscle and the **dartos muscle** composed of smooth muscle. These muscles are responsible for lowering and raising the scrotum to maintain an optimal temperature in the testes for spermatogenesis.

Sitting atop the testes is the **epididymis,** a tightly coiled tube where sperm are stored for about 60–80 days. Extending from each epididymis is the **vas (ductus) deferens,** a tube that connects to the ejaculatory ducts inside the pelvic cavity. The vas deferens along with blood vessels, nerves, and lymphatic vessels creates a structure called the **spermatic cord.**

During the emission process, the matured sperm move from epididymis into the vas deferens. The paired vas deferens direct the sperm into the pelvic cavity towards the **seminal vesicles.** The two **seminal vesicles** are tightly coiled, sac-like glands found posterior to the bladder. These glands produce a viscous, alkaline fluid that provides nutrients and protective chemicals for the sperm; these secretions constitute 60–80% of the liquid portion of semen. The secretions from the seminal vesicles travel through seminal ducts and converge with the vas deferens at the prostate gland to form the **ejaculatory ducts.**

The **prostate gland** is a doughnut-shaped, golf ball-sized gland located inferior to the bladder. It produces a slightly acidic liquid that provides protection to the sperm. Inside the prostate, the two ejaculatory ducts unite with the urethra to form the **prostatic urethra.** This is where the reproductive system and the urinary system meet to form a common route. Once beyond the prostate, the urethra is known as the **penile urethra (spongy urethra).**

After the prostate gland, the paired **Cowper's glands (bulbourethral glands)** contribute the last liquid component of semen. This fluid is a thick, clear, alkaline mucus secreted into the penile urethra.

The mixture of sperm and the fluids released by the accessory glands mentioned above is called **semen.** During ejaculation, **semen** travels through the penile urethra to the glans penis where the urethral orifice (meatus) is located. This is where the semen exits the body. Because the urethra serves both the reproductive and urinary systems, there are mechanisms that ensure the urethra can only be used by one system at a time. The urinary portion of the urethra contains internal and external urethral sphincters which are normally closed. These do not open until the bladder is full, and the micturition reflex is activated.

Required Reproductive System Structures for Practical

After completion of the lab, you should be able to identify and correctly spell the following reproductive system structures on the available models.

Female Model

- Labia majora
- Labia minora
- Ovary
- Vagina
- Uterus
- Fimbriae
- Fallopian tube (oviduct)
- Round ligament
- Clitoris
- Cervix

Male Model

- Testicle
- Epididymis
- Glans penis
- Shaft (body) of penis
- Spermatic cord
- Vas deferens
- Seminal vesicle
- Prostate gland
- Bulbourethral (Cowper's) gland
- Corpora cavernosa
- Urethra
- Urethral orifice (meatus)

35

Organ System Dissection

Your lab instructor will provide an overview of the organs of several organ systems in the human cadaver. The specific organs that may be examined will vary based upon the particular dissection; therefore, not all organs that are discussed below may necessarily be viewed. Based upon available time, you may more closely examine the organs in small groups.

A | Organs of the Respiratory System

In the neck area, find the ***larynx,*** or voice box. If a midsagittal incision has been made, observe the ***vocal cords*** contained within the larynx.

Examine the ***trachea*** and notice the cartilage rings that keep the trachea open at all times; collapse of the trachea would be fatal.

Follow the trachea into the thoracic cavity until you see where it bifurcates (divides) to form the ***bronchi***—one for each lung. The bronchi enter the lungs at a medial opening known as the ***hilus*** (or ***hilum***). Also entering and exiting the hilus are the ***pulmonary arteries*** and ***veins.***

Before examining the lungs, note the ***pleural membranes*** which cover them (the ***visceral*** portion) and which line the thoracic cavity (the ***parietal*** portion). Between the two pleura is the pleural space which is filled with ***serous fluid.***

The lungs are divided into ***lobes*** by ***fissures.*** The right lung has three lobes while the left has two.

Between the lungs is the ***mediastinum*** which contains the heart, esophagus, trachea, thymus gland, and the great vessels associated with the heart.

Even though you have previously dissected a heart, take the time now to re-examine the ***pericardial sac,*** as well as the external anatomy of the heart.

B | Organs of the Digestive System

Upon examination of the ventral aspect of the abdomen, you will notice some fatty tissue. This is the ***greater omentum*** which covers the abdominal organs. (The ***lesser omentum*** will be found later.)

The digestive system is composed of organs of the ***gastrointestinal (GI)*** or ***digestive tract*** which extends from the mouth to the anus, and the ***accessory organs*** which are organs that lie outside the GI tract but are important in contributing digestive enzymes and other materials required for digestion.

The Digestive Tract Organs

Beginning with the thoracic cavity, locate the esophagus behind the trachea. The ***esophagus,*** composed primarily of smooth muscle, will appear collapsed since it does not contain food.

Follow the esophagus through the diaphragm and into the abdominal cavity. Just inferior to the diaphragm the esophagus leads directly into the ***stomach,*** which lies slightly to the left of the midline.

Using your textbook as a reference, locate the following parts of the stomach. The ***fundus*** lies above and left of the entrance of the esophagus. The ***body*** is the central portion of the stomach. The portion of the stomach which connects to the small intestine is the ***pylorus.*** It contains a ring of circular smooth muscle (sphincter) which is known as the ***pyloric valve*** or ***sphincter.*** This valve promotes the movement of food one way—from the stomach into the intestines. Examine the interior lining of the stomach. If the stomach is empty, it is arranged in folds known as ***rugae.*** If the stomach were distended with food, the rugae would be stretched flat.

The ***small intestine*** is comprised of three parts. The first portion of the small intestine is the ***duodenum,*** the primary area for digestion and absorption. The duodenum is approximately 10 inches long.

Cutting lengthwise through the duodenum, you may feel the velvety lining. The lining of the small intestine is covered with millions of microscopic ***villi,*** which increase the surface area of the lining of the small intestine 600 times (enough to cover the surface of a tennis court). The increased surface area allows for more efficient digestion and absorption.

Next is the ***jejunum,*** which is followed by the ***ileum.*** The jejunum is approximately 8 feet long and the ileum almost 12 feet long. While they appear to look the same when observing gross anatomy, they are different histologically.

The ileum is the terminal portion of the small intestine and enters into the ***large intestine*** at a pouch-like area called the ***cecum.*** The ***appendix,*** if present, is a very small tube extending from the cecum. It serves no digestive role.

The ***large intestine*** is larger in diameter than the small intestine, thus its name. Extending upward from the cecum on the right side is the next portion of the large intestine, the ***ascending colon.*** This crosses to the opposite side (from right to left) as the ***transverse colon,*** then the ***descending colon*** (along the left side), the ***sigmoid colon,*** and finally the ***rectum*** and ***anus.*** Also in the large intestine are sac-like pouches called **haustra.**

Accessory Organs

If exposed, examine the three pairs of ***salivary glands*** in the neck and jaw area. These glands, the ***parotid, submaxillary,*** and ***sublingual,*** produce saliva, which aids in the digestion of food as it passes through the mouth.

The ***pancreas*** is a flat, spongy organ located near the underside of the stomach. Part of the role of the pancreas in digestion is to produce several digestive enzymes and neutralize the stomach acids. The pancreatic juices enter the duodenum through a ***pancreatic duct.***

The ***liver*** is the large, dark-colored organ in the upper right quadrant of the abdomen. As a digestive organ, the liver extracts digested nutrients from the blood of the hepatic portal system. It also produces ***bile*** which aids in the digestion of fats.

The liver has four lobes. Under the right medial lobe, you will find the ***gall bladder*** which stores the bile that was produced in the liver. Occasionally, ***gallstones*** are found within the gall bladder. These hard stones are crystallized bile salts, minerals, and cholesterol that can form when bile is too concentrated.

Between the liver and the duodenum is the ***lesser omentum.*** Intertwined within it is the ***common bile duct.*** The greenish common bile duct is formed by the joining of the ***hepatic ducts*** from the liver and the ***cystic duct*** from the gall bladder. The common bile duct joins with the pancreatic duct from the pancreas and enters the duodenum.

C | Organs of the Urinary System

Beneath (posterior to) the digestive organs, you will find the ***kidneys.*** The kidneys are located behind the peritoneum and are, therefore, referred to as ***retroperitoneal.*** The right kidney is lower than the left to accommodate the liver.

Three layers of tissue surround the kidneys. The (1) ***renal fascia*** anchors the kidneys to the dorsal body wall. The kidneys are embedded in a protective and insulative layer of fat known as the (2) ***adipose capsule.*** In addition, each kidney will be covered by a membrane known as the (3) ***renal capsule.***

Take a moment to review the relationship of the ***renal artery*** and ***renal vein,*** which enter and exit the kidney at the medial opening known as the ***hilus (hilum).***

Also leaving the kidneys at the hilus are the ***ureters,*** tubes which carry the urine from the kidney to the bladder. The ureters are also considered to be retroperitoneal.

Follow the ureters inferiorly to the ***urinary bladder.*** The bladder walls are composed of smooth muscle while the lining has a specialized type of stratified epithelium, transitional epithelium. You will recall that this type of epithelium has rounded cells near their free surface, which are stretched flat when the bladder is full. Urine is stored in the bladder until it leaves the body via the ***urethra.***